NOTE ON TH

Heine Bakkeid grew up in the rugged landscape ...
Norway. *I Will Miss You Tomorrow* is the first instalment in a
new crime series, highly acclaimed by critics in his home country, and has earned him a reputation as a virtuoso of darkly
atmospheric suspense.

NOTE ON THE TRANSLATOR

Anne Bruce graduated from Glasgow University with degrees in Norwegian and English. She lives in Scotland.

I WILL MISS YOU TOMORROW

HEINE BAKKEID

Translated from the Norwegian
by Anne Bruce

RAVEN BOOKS

LONDON · OXFORD · NEW YORK · NEW DELHI · SYDNEY

RAVEN BOOKS
Bloomsbury Publishing Plc
50 Bedford Square, London, WC1B 3DP, UK

BLOOMSBURY, RAVEN BOOKS and the Raven Books logo are
trademarks of Bloomsbury Publishing Plc

First published in 2016 in Norway as *Jeg skal savne deg i morgen* by Aschehoug

First published in Great Britain 2019
This edition published 2020

A catalogue record for this book is available from the British Library

ISBN: HB: 978-1-5266-1077-5; TPB: 978-1-5266-1076-8; PB: 978-1-5266-1074-4;
EBOOK: 978-1-5266-1075-1

2 4 6 8 10 9 7 5 3 1

Typeset by Integra Software Services Pvt. Ltd.
Printed and bound in Great Britain by CPI Group (UK) Ltd, Croydon CR0 4YY

MIX
Paper from
responsible sources
FSC® C020471

To find out more about our authors and books visit www.bloomsbury.com
and sign up for our newsletters

*There are days when you don't see that the
mountains are still standing.*

*Hours come where everything is litter at high-water
mark,*

all along the black shores.

Times that you don't recognise anyone.

– Herbjørg Wassmo

Seven minutes

At three minutes past five, regret hits me: a tidal wave rushing through my body as I wheeze and gasp in panic. I tremble and shudder and thrash my legs to break free, but it's futile.

Two minutes later and my lungs' convulsions finally stop. I notice I no longer need oxygen as I just hang there on the rope, feeling my system shut down, one body part at a time.

Eight minutes past, and I can hear the water drumming against the floor tiles beneath me. A rasping noise leaks out from my throat as steam and tears slide down my cheeks, sluicing what is left of me down into the drains on the floor. I'm freezing.

Then she is standing there, right in front of me, as grey as the rest of the room. I feel an impulse to laugh, to shout for joy at seeing her again. I try to open my mouth to tell her this, to say that I'm as happy as anyone could ever be. Instead I hear a crack and the very next moment I'm lying on the floor. The water from the prison showers is washing over my face as the minute hand on the wall clock jerks forward another notch.

Ten past five.

WEDNESDAY

CHAPTER 1

Stavanger is the city covered in dog shits. They stick up like small mushrooms and, in a grim way, colour the place in a palette of browns.

Stepping over yet another, I hurry along Pedersgata towards the city centre. Originally, Aetat, the job centre, was based in a ground-floor, open-plan office overlooking Klubbgata and the Breiavatnet Lake. Its large windows meant passers-by could look in and look down on the poor souls in there, those desperately trying to hide themselves from anyone they knew behind plastic potted plants, room dividers and lampshades, while explaining why they were no longer employed. Since I was last in the city, this stage-set for societal failures has changed its name and address and moved into more traditionally furnished premises nearby.

I pull a queue ticket from an automatic dispenser and sit down on a red sofa in a room with no windows or oxygen, the odour of perspiration, sweaty feet and general stink of losers oozing from the clientele filling your nostrils the moment you cross the threshold. Despite all the people thronging around me, there's hardly a sound to be heard. Only a buzzing fly and sporadic keystrokes break the silence.

'Thirty-eight?'

A female adviser pops her head out of an open door to scan the waiting room and beckons me to approach. As soon as I'm close enough she gives me a limp handshake.

'I'm Iljana,' she says, her eastern European accent unmistakable as she settles into her office chair. 'Please take a seat.'

'Thanks,' I reply, and do as she asks.

Iljana's wispy, dark brown hair is pinned up at the back. She's wearing a grey suit with big black buttons, the kind used for eyes on ancient teddy bears. 'May I take your date of birth and personal ID number?'

Once I have given them, Iljana turns away and types.

'Thorkild Aske?' she asks.

'That's me.'

'How can I help you, then? Have you registered as a job-seeker before?'

'No.'

I hand her the letter I've brought from the social worker at Stavanger Prison.

Iljana leans forward to the desk and computer screen while she reads. Once finished, she gives a vague smile. She has small teeth, unnaturally so, almost like a child's, and her eyes are as grey as her suit.

'Well then, Thorkild.' She folds her hands on her lap. 'The prison social worker writes that you have agreed to take part in an interdisciplinary project to facilitate your return to society after serving your sentence. And that's good, of course.' She emphasises the word *good* and smiles again.

I nod.

'I've attended a meeting with Correctional Services who found me somewhere to live, allocated me a psychiatrist and GP and talked to various assessment units. They've all helped me form a support team enabling me to discuss the past and plan the future in order for me to break away from my criminal career. If you ask me, I'd say I'm close to a hundred per cent rehabilitated.'

Unable to see the humour in what I'm saying, she returns to the computer screen. 'You trained as a policeman.'

She scrolls down the screen as she talks. 'Sergeant, Inspector, Chief Inspector, assigned to Internal Affairs and subsequently employed there on a permanent basis.' Hesitating, she runs the tip of her tongue along those tiny teeth before turning to face me once again.

I beat her to it. 'Police officers who go after police officers.'

'I see.' She nods. 'Then it would be logical for us to look for a job in the police when you're ready, don't you think?'

I return her smile. 'Dismissed from the service,' I say, aware that the pains in my cheek and diaphragm are surfacing again. At the same time, my mouth is so dry it's difficult to speak.

'Sorry?'

Unscrewing the cap from the bottle of water I've brought with me, I drink, hoping the water will make everything better. 'I was also sentenced to loss of my police career.'

'For how long?'

'For the rest of my life,' I answer, replacing the cap and setting the bottle down on the floor beside the chair. The tingling inside my face is changing into a pulsating, aching throb. 'And a bit longer.'

'So what were you thinking of doing now, then?'

'That's what I was hoping you could tell me.' I retrieve the bottle of water and hug it between my hands. The pain, the smell, the light, the shortage of oxygen in the air and the obligation to sit here talking to a stranger, yet another official, makes me restless. I long to be alone in a room with no reflective surfaces. Yet I know that I have to endure this to emerge on the other side. Ulf says there's no way round it.

Iljana glances down at the letter again before returning to the screen. 'It says here that you wish to apply for work-assessment allowance while you are receiving medical treatment?'

I nod. 'There's still some disagreement about what degree of fitness I have for work following my' – I draw quote marks in the air – '*accident at work*. But my contact officer, my prison social worker, the hospital staff, the prison chaplain, my GP, psychologist and psychiatrist and I have all reached the conclusion that I should make a dignified effort to get back to work when I can.'

'Accident at work?'

'Doesn't it say that as well?' I point to the letter. 'This individual – Thorkild Aske – tried to hang himself from a pipe in the communal showers several months into his prison sentence. In the middle of the February half-term holiday, into the bargain.'

'What happened?' she croaks.

'The rope snapped.'

Iljana stares at me as if she's scared that any minute now I'll attack her with one of the plastic bananas in the fruit bowl on her desk. 'Well,' she ventures tentatively, taking a deep breath and clearing her throat. 'Are you thinking of retraining for something else?'

'For what, though?' I squeeze the sides of the bottle so tightly that water drips over my fingers and down on to the floor. 'A forty-four-year-old brain-damaged petroleum engineer? Stockbroker? Dental assistant?'

Iljana steals a glimpse at the clock in the right-hand corner of the screen before announcing with fresh resolution: 'I suggest we await the outcome of your application for work-assessment allowance. In the meantime, we'll explore possible avenues for returning to the world of work, outside the police force.' She types on the keyboard again, scrolling down and typing some more until, finally satisfied, she turns to face me. 'What do you think about working in a call centre?'

CHAPTER 2

In an effort to banish the pains in my cheek and belly, I decide to eat. I buy a sandwich in a café close by the Work and Welfare Office, and afterwards set off up Hospitalgata and walk on through Pedersgata in the direction of the bedsit that I've been allocated by Correctional Services.

In the mailbox I find a furniture catalogue and a letter addressed to me. I know what's inside the envelope. They're always the same; the only thing that changes is the children's ages. They grow up, even though the faces are never the same. When I received the first one, they were pictures of babies taken from magazines and catalogues. In the beginning, she also sent cuttings of cots, rattles, feeding bottles and breast pumps.

Bringing the letter and the catalogue with me, I go up and let myself into the bedsit. I put the post on the table between the sofa and the TV table minus a TV, before crossing to the kitchenette and taking down my medicine dispenser from the cabinet above the hotplate. I open the Wednesday compartment, tip the contents of the middle space into my palm and wash them down with a mouthful of water. I switch on the coffee machine and then settle down on the sofa with the letter.

This time there are two cuttings inside the envelope. One is of a boy of about seven or eight with wavy brown hair, wearing a colourful T-shirt with a printed design of a fish in a cap and snorkel swimming around a reef. The text underneath reads: *Find clothes made for fun and games – jeans, trousers, T-shirts, jackets and lots more. Our colourful, hard-wearing clothes suit all children.*

The next cutting is a girl of the same age. According to the text below, she is dressed in a powder-pink jacket with a detachable fake fur collar, tight denims and a matching T-shirt. *We have everyday jeans, practical clothing for play-time, party favourites and everything in between …*

I tuck the cuttings carefully into the envelope and shove it to the furthest end of the table with the furniture catalogue, before reclining on the sofa and closing my eyes.

Then the phone rings.

'Well?' asks a gruff male voice with a distinct Bergen accent, drawing on a cigarette with a greed that's close to intimacy. Ulf Solstad is a psychiatrist by profession and leader of my newly established support team. 'How did the meeting go?'

I made Ulf's acquaintance during my stretch in Stavanger Prison, where he was serving a sentence of eight months for extortion, though that does not appear to be deterring his clientele now. In fact, he is even more sought after by city suits with money and problems than before his stint inside.

'Superb,' I answer sourly. 'A promising career is predicted for me in the call-centre sector.'

'Take it easy.' Ulf prolongs his vowels to an abnormal degree, even for someone from Bergen. 'Just persevere and follow the rat runs through the maze. The bureaucrats set them up precisely for people like you, to sift out the ones who aren't strong enough. I promise, as soon as we've got

11

you over the hurdle of the work-assessment allowance, it won't be long until you're permanently assigned to the ranks of the unemployable. And in the meantime, wait for your benefit cheque.'

'What?'

'Listen,' Ulf breaks in as I struggle up from the clutches of the sofa in search of my water bottle. 'I'm honoured that you want to have me as a member of your support team, and promise to do my best to ensure you achieve the life you desire for yourself, Thorkild.' I can hear him puff out cigarette smoke.

'I need more Sobril.' I snatch the bottle that has fallen on the floor and rolled partway under the two-seater. 'I'll soon have none left. Besides, we'll have to increase the dose of OxyContin.'

'Has the pain got worse?'

'Yes,' I tell him. 'And I've started getting cramps in my legs when I walk.'

'Maybe we should look at the dose of Neurontin instead?'

'No!' I snap, pressing my index finger into the damaged tissue on my cheek. Soon my whole face will be on fire with pain. 'It gives me a headache. And so does Risperdal. I can't tolerate them.'

'Thorkild, we've talked about this before. Neurontin is specifically for nerve pain, and you'll probably have to take that for the rest of your life. Risperdal is an antipsychotic that you absolutely do still need. It's always the benzodiazepines people think they need more of, because they reduce anxiety, those and the Oxys. That's always the case, but they're addictive, as you well know. If we're going to reduce the doses, we'll start with them, and then we can eventually weigh up whether you've found your feet now that you're out of prison. What do you think?'

'I can't sleep,' I grumble, using my heel to push the mail over the edge of the table. I know he's right, and it pisses me off.

'Yes you can,' Ulf answers, unruffled. 'That's why I've given you Sarotex.' He coughs violently before continuing: 'You are still taking them all, aren't you?'

'What do you mean?'

'Your medicines? You're taking them?'

'Of course.'

'The Risperdal as well?'

'Yes.'

'You know you need them, Thorkild.'

'Yes, I know that,' I answer, far too quickly.

'Get a grip!' Ulf interrupts me. 'I'm not your damned hunched-up chaplain trying to fill his quota for heaven.' He's breathing heavily again. I've upset his smoking ritual and he's going to have to light another one as soon as this one's sucked all the way down to the filter.

'He said I was a bee with no flowers.'

'Who?'

'The prison chaplain.'

'Are you kidding?'

'No.'

I hear the flick of the lighter as Ulf starts his next cigarette. 'Tell me the story, Thorkild. Can you do that for me?'

I resolve to make this smoke a good one and relate the story. 'I'm a bee in a world without flowers, and it's up to me to decide how I want to spend my time until winter comes.'

'Winter?' Ulf exhales with gratified ease. I can hear him through the receiver, the pleasure as he puffs in and out.

'Winter touches all of our lives, sooner or later,' I plough on, feeling my muscles tense. I sink back into the sofa, letting it suck me down into the cushions. The effect of the powerful pills causes the pains to disperse and vanish.

'You're kidding me, aren't you? Tell me you're pulling my leg, Thorkild!'

'No. I'm not kidding. It's like waves breaking against rocks.'

'That's the worst thing I've ever heard.' He pauses. 'Can I use it?'

'Be my guest!'

'Thorkild,' Ulf says, abruptly, decisively, as I'm about to hang up. 'There's someone who wants to talk to you.'

'Oh?'

'Someone you know. From before.'

He hesitates, as if he still hasn't quite made up his mind whether this is something he really ought to tell me before the support team in its entirety has had a chance to analyse the matter.

'Who?'

'Frei's uncle,' Ulf finally says before adding, 'and his ex-wife, Anniken Moritzen.'

'Arne Villmyr?' I feel the anxiety come creeping back. All at once my mouth is dry as dust and the sunlight slanting round the blanket stretched over the window is hurting my eyes. 'Why?'

'It's not to do with Frei,' Ulf answers in a strained voice, as if still not entirely sure about what he's doing. 'Arne and his ex-wife, they have a son—'

'Arne's gay,' I break in. I'm not keen on the direction this conversation is taking, and at the same time I feel a mounting disquiet inside me and yearn to just hang up and shut it all out.

'All the same,' Ulf says calmly, giving me no excuse to disconnect, 'he has an ex-wife and a son—'

'What's that got to do with me?' I squeeze my eyelids shut and twist my face away from the piercing light.

'If you'd just let me finish,' Ulf continues, before exhaling loudly. 'You see, Anniken Moritzen is one of my patients. She needs ...' Again he hesitates, taking a deep breath before he continues: 'They need help. Their son has disappeared.'

'I'm not a private investigator.'

'No, God forbid,' Ulf wheezes. 'But Anniken is a friend and I don't know how much I can do for her in this situation. Besides, you and Arne have this shared past, one that you can't escape, no matter what, and now he's asked to be allowed to speak to you. I think you owe him that much?'

The pain is pressing against the skin on my face, my eyes and the outer layer of my brain. 'Please,' I groan through gritted teeth. 'Not today, not now.'

'Talk to them. Hear what they have to say.'

'I don't want to.'

Ulf sighs again. 'You've drawn your lot, Thorkild, gone down into the cellar and come back up again. A changed man.' He gasps for breath as he stubs out his cigarette. Half-smoked, spoiled. 'Don't let that bedsit become your new prison cell. You need to get out, talk to people and find out who you want to be in this new life of yours, beyond the prison walls.'

'I know,' I whisper, sinking into the sofa again. I open my eyes, force my gaze to confront the glowing light that fills the room, and hold it there until my eyes are brimming.

'What did you say?'

'That I know.'

'Sure?' Ulf Solstad drops his voice to a more therapeutic level. 'OK,' he says when I don't respond. His breathing is quieter now. 'Then you can pop in afterwards, and we can take a look at those medicines of yours at the same time. Will you do that? Will you?'

Ulf Solstad's third attempt to smoke the perfect cigarette would have to be made in splendid isolation.

CHAPTER 3

Arne Villmyr stands beside Anniken Moritzen, who is seated in an office chair with her hands folded on the desk in front of her. Behind them, three floor-to-ceiling windows frame the Forus landscape with its roaring motorways and industrial buildings. Arne is dressed with the same good taste as the first time I met him almost four years ago in his villa at Storhaug Vest, but his hair has grown thinner and his face paler.

'Thorkild Aske?' Anniken Moritzen enquires, without rising from her chair.

'Yes,' I reply, and take a reluctant step closer.

'Nice to meet you,' she answers, joylessly. I detect apathetic contempt as she finally offers me her hand to shake: I observe that the right-hand corner of her mouth refuses to respond to the smile impulse and remains paralysed. The outcome looks more like a sneer than anything else.

Arne Villmyr makes no move to return the gesture when I stretch out my hand to shake his.

'I have a picture of him.' Anniken Moritzen produces a photograph from one of the desk drawers.

'That's nice,' I venture, leaning towards her and using both hands to pick up the photograph so that it doesn't slip between my fingers and fall to the floor.

'It was taken five months ago when we visited my parents in Jutland.' Anniken speaks in a sort of upper-class accent, making no secret of her Danish origins. She's in her mid-fifties, I'd guess, dressed in a dark blue jacket and matching skirt, with a white blouse, its two top buttons left open. It strikes me that she must be a whole head taller than her ex-husband.

'It looks like a great place for a child to grow up.'

She looks at me as if about to say that she knows what I'm getting at, but lets it pass. 'It's the last picture I have of him.' She lingers over the image, which shows her in her parents' garden, barbecuing food and drinking lemonade. Her son Rasmus is tending the barbecue, dressed in a pair of red Liverpool football shorts, flip-flops and a white chef's hat. He is tanned and has an athletic build. His grandfather is toasting the unseen photographer with a dram while Anniken Moritzen sits smiling in a chair.

'Rasmus and some of his school friends spent the past year travelling round the world in a yacht.' Anniken looks dreamily at the back of the photo as she talks, as if to soak the last dregs of energy out of the memory it prompts. 'But after a trip to northern Norway, he hit upon the idea of turning a former conference centre in a lighthouse there into an outdoor activity hotel.'

'Activity hotel?'

'Wreck diving, harpoon fishing and so on. Rasmus says it's very popular overseas.'

'How old is he?' I ask, even though I already know. On the bus to Forus, I found a report in an online Tromsø newspaper about a missing 27-year-old man, assumed to have perished in a diving accident not far from the village of Skjellvik in Blekøyvær district.

18

'Our Rasmus is twenty-seven.'

'And when did he go there?'

'Anniken bought the place for him to use last summer,' Arne says. Behind him, the sea breeze has begun to tug at the rain clouds. Silvery shades of grey scud in a south-westerly direction.

Anniken nods without looking at either of us. 'The entire little island and its lighthouse have been left abandoned since the former conference centre went bust back in the eighties. Rasmus went up immediately afterwards with a few of his friends to help him with the renovations until the holidays were over.'

'When did he go missing?'

'The last time I spoke to him was on Friday, five days ago. The police found his boat yesterday morning, so they believe he went out diving on either Saturday or Sunday.'

'And you?' I glance up at Arne Villmyr, who is staring vacantly ahead, like a soldier at attention, while the rain has begun to tap against the windows at his back.

Arne shakes his head gently just as the heavens open above us with a resounding crash and sheets of water start to pour down the glass.

'They don't have much contact,' Anniken answers, pressing her arms to her sides, as if finding herself suddenly out there in the downpour.

'Was he alone when he disappeared?' I ask, my gaze shifting to the grey tones beyond the windowpane.

'Yes, for the past month he's been there on his own.'

'Why do the police believe he drowned?' *A bit more, Thorkild,* I intone to myself, deep down. *Just a few more questions, and then you can go home.*

'When they found the boat, the diving gear was gone. Rasmus was in the habit of going out to the skerries beyond the lighthouse to dive when he had spare time. On Friday he said he wanted to go out diving that weekend if the weather was good enough.'

'Have you any reason to think anything else might have happened to him, or that this isn't a diving accident?'

'No.' Irritation is etched on her face. Probably I'm interrupting her at the same point as everyone else she has spoken to since her son disappeared.

I feel an impulse to go over and give her a shake, tell her she has to wake up, stop dreaming. It's leading her nowhere. These dreams we dream with our eyes wide open.

'I went up there as soon as I got no answer. I felt something was wrong.' Anniken Moritzen turns to face her ex-husband. 'I told you that, didn't I? He would have returned my call. He always phones back.'

Arne puts his hand carefully on her shoulder and gives a silent nod.

'But the weather was stormy up there,' she continues. 'The local police chief and his assistant refused to take me out to the lighthouse and treated me like a hysterical nuisance they could just show to a hotel room in Tromsø, a hundred kilometres away, while they went on sitting there in their offices without lifting a finger. No one would help me, no one would do anything. They're just sitting there, do you understand? They're just sitting there doing nothing while my boy is somewhere out on the sea, needing help!' She sobs bitterly. 'That's why I came home again, Arne,' she whispers, her eyes brimming with tears. 'Because you said you'd find someone who could help us. Someone who would listen. Do you remember that? You promised to find somebody who could help us.'

Arne closes his eyes and keeps them like that as he nods, over and over again. Anniken Moritzen turns to me once more. 'You, Aske.' She takes a deep breath and wipes her cheeks with the back of her hand. 'They'll talk to you. I know that. You'll be able to find him,' she says, smiling warmly at what this fantasy conjures up inside her mind. Hugs to herself the illusion that there is still time. 'Yes, you'll be able to find Rasmus for me.'

Once again I lower my gaze to the man in the photograph. When I was the same age as Rasmus, I was a chief inspector of police in Finnmark and spent my time dissuading drunken snow-scooter drivers from shooting road signs in the area to smithereens. 'I'm not a detective,' I begin to say, putting the photograph down on the desk in front of me.

'We'll pay you,' Arne Villmyr interjects. 'If you're worried about the money side of things.'

'It's not that—' I'm going to say that it's too late. That no one can disappear at sea under such circumstances and then return again almost a week later. But Arne Villmyr has already let go of the chair's back and is on his way around the desk.

'Come on,' he says, grabbing my upper arm. He points brusquely at the door. 'Let's take this conversation outside.'

We leave Anniken Moritzen and step out into the corridor, all the way along to the lift.

'So,' he says, releasing my arm when we get there. He presses the lift button and turns to face me. 'Now it's just the two of us again.'

'Listen—'

'My son is dead,' Arne Villmyr says calmly as he adjusts his tie. 'There's nothing to investigate,' he continues, when he's finished fiddling. He looks at me. 'What you have to do is go up there, find his dead body and bring it home.'

'My God,' I exclaim, spreading my arms in dismay. 'How am I supposed to do that?'

'Swim, dive, jump through red-hot rings of fire, I don't give a shit how you do it. I lost Rasmus when I left my family years ago. But he can't just vanish as if he has never existed. We need a grave to visit.' Arne's jaw muscles tense and his eyes harden. 'And I've convinced myself you're the person who can give us that. Call it payback for an old debt, call it whatever the fuck you like, just find him.'

'Arne,' I venture. 'Please. What happened with Frei, you can't use that, now. Not in this way—'

'That's enough, Thorkild,' he says, just as unruffled and steady as before, although the tightness in his jawline is still there. 'You're not to speak about her,' he goes on. 'Not yet. Not until you've found Rasmus and brought him home again. Afterwards, you can crawl back down into the hole you came from and do whatever you like with the rest of your life. But until then: you search and I pay. Got that?'

The lift has already come up and disappeared down again when Arne turns to stalk back to Anniken Moritzen's office. He stops in front of the door with his back to me. 'Give us a grave, Aske,' he says as he puts his hand on the door handle. 'Yet another grave. Is that so damned much to ask?'

CHAPTER 4

My bedsit is always at its greyest in the evening. The faint streetlight filtering in only emphasises the deathly pallor that fills the room. Outside I can hear rain trickling through the gutters and the roar of cars driving to and fro across the suspension bridge linking the city and Grasholmen, Hundvåg and the islands beyond.

I am lying on the sofa. In the background, the radio is crackling in time to the gurgle of the coffee machine, Leonard Cohen's hoarse voice whispering out. Early on in my course of treatment, Ulf suggested I should listen to music after my evening dose to soothe my brain against insomnia.

Rolling over on to my side, I turn my face to the darkness and the chair between the wall and the kitchen corner, and hear a sound from there. 'Frei,' I gasp, and surface at the same moment that Cohen's voice sounds again like a melodious cello chord.

The room smells suddenly earthy and dank. I crawl cautiously off the sofa and get to my feet. Inside my body, I feel an acute tingling begin. Anticipation, delight at what is about to occur.

I cross to the chair and stretch out my hand to the darkness before me at the same time as the radio crackles once again, before the music fades, to be replaced by a rasping background noise that blends with the autumn rain outside.

We dance, close in tune with the hum of the fridge in the cramped bedsit kitchenette. No music, no light, just the murmur of the rain and the fractured sky above us. I no longer notice my cheek burning with pain. Our bodies sway gently from side to side.

'I never thought I'd see you again,' I sob convulsively, and the tears in my damaged tear ducts finally break free.

Her wild blonde hair has lost all its colour and gloss. The scent of unknown plant extracts, herbs and vanilla is gone, replaced by a raw stench of sterilising soap and cold earth. The fragrance of her, of us, has vanished, washed away by the time we have spent apart.

'But you're back.' I hold her fingers in mine, draw her body closer and try to bury my face in her hair, inhaling again and again before her head falls heavily on my chest. 'Come,' I sigh with exhaustion, encircling her waist with my arm, and pull her body closer to mine.

We stagger across to the sofa bed, where I haul back the blanket and drape it like a cape over my shoulders before climbing in. I notice I am shivering as her cold body meets mine.

Shivering with happiness.

CHAPTER 5

My first day with Frei, Stavanger, 22 October 2011

I had recently returned to my job in Internal Affairs in Bergen after almost an entire year in the States. Since I'd been issued an assignment in Stavanger, I found myself on the steps outside a villa on the western side of the city's Storhaug district, where I was to meet a lawyer connected with one of two cases I was in the city to investigate.

The first of these concerned a possible breach of professional secrecy, allegedly committed by a court employee in connection with a compensation case involving two foreign oil companies based in the city. The other was far more serious. A police officer at Stavanger police station had been reported by a colleague for a number of violations of the criminal code and the law relating to weapons. I had arranged an interview with the accused officer later that week.

'Who are you?' a voice behind me suddenly enquired as I made to ring the doorbell. The sun was shining and the air mild, although autumn had already left its mark on the leaves in the trees. I wheeled around. Her eyes were narrow and bright and her oval face crowned by a halo of curls.

'Thorkild Aske,' I answered, taking a step to one side. 'And who are you?'

'Frei,' she said, walking up the steps to stand beside me. She was in her early twenties and looked as tall as me. 'What are you doing here?'

'I have an appointment right now, at five o'clock, with Arne Villmyr. Does he live here?'

'Are you from the police?'

'In a way.'

Frei put her hand on the railing and leaned back as she scrutinised me with a passionate, youthful gaze that made me want to turn away in shame at my own age and physical inadequacy.

'In what way, then?'

'I work for Police Internal Affairs, we're the ones who ...'

She shot me a crooked smile before I managed to finish my sentence, and pushed the door open with her free hand. 'So what are you actually doing here at Uncle Arne's? Is he going to be arrested?'

'As I said, I—'

'Frei? Is that you? Hurry on in and close the door behind you,' a man's voice sounded from inside the house. 'I think we've got a wasps' nest under the verandah, and I don't want those horrible creatures crawling into the living room.'

My eyes glanced up at the sky before coming to rest on Frei again. She had kicked off her sandals in the hallway and was sauntering barefoot to the living room at the end of a bright corridor.

'There's a man here,' she said, just loudly enough so that I could hear her. 'Some kind of policeman who wants to talk to you ...'

THURSDAY

CHAPTER 6

The sight that greets me in the mirror in the mornings looks like a spectre from the underworld. My complexion is wan and grey from lack of sunlight and vitamins. My eyes are shrunken, with curved purple semicircles beneath and swollen eyelids that never open more than halfway.

I wash my face and run wet fingertips over the half-moon-shaped scar below my eye, following the line down to the ragged pattern in the middle of my cheek, touching every single crater and blemish. The pain strikes almost at once.

'I can't,' I whisper to the face in the mirror as I fumble with the pack of individual doses that contains my morning medicine. 'He should have known better. I'm not ready.'

After I've taken my pills, I get dressed and cross to the window, tugging back the blanket-curtain to look out. It's one of those days – the sun not shining, and everything a pale bluish-grey, as if the light in the sky refuses to switch on.

I'm about to turn away when I catch sight of a man in a helmet wearing a snug T-shirt and cycling shorts making his way towards the bedsit building. Stopping outside the entrance and getting off his bike, he glances up at the window where I'm standing, and takes out his mobile phone. Ulf Solstad, about six foot four in height, is powerfully built and his head is almost completely

29

bald, but for a ponytail of thick, red hair, tied in a kind of samurai style.

I drop the corner of the blanket and retreat to the sofa just as my phone starts ringing.

'Good morning, Thorkild,' Ulf says, sounding slightly out of breath, when I finally answer. 'Anniken Moritzen phoned me a short time ago. She says she's received a message from you.'

'Yes.' I sink on to the sofa and try to focus on the tingling in my cheek, to push it to the front of the pain queue and let it take control of the moment. 'I can't go.'

'Why not?'

'There's no point.'

'Because?'

'Arne says their son is dead.'

'He's probably right.'

'My God,' I groan in despair. 'So what the hell is it you all expect of me?'

'We're doing this for Anniken,' Ulf replies calmly. 'One day they're going to find their boy, bloated and rotting from his time in the sea, where fish and crabs have been nibbling at him. But it's still her child, you understand? And I'm telling you: she's in no state to cope with what's ahead. You know the police lingo and routines in such situations and how things work. Maybe most of all, this is a way for her to demonstrate to herself that she's not giving up. No one can give up until they know for certain, Thorkild. Until all avenues have been explored. Don't you agree with me?'

I don't say anything, just sit there with mobile in hand as I stare at the blanket covering the window.

'Come down, Thorkild,' Ulf adds when I don't respond.

'No.' My voice is breaking, and tears are pressing at my eyes, without finding release through my messed-up tear ducts.

'Let me in, then, and I'll come up.'

'I don't want to.'

'I'm not leaving until you come down, or else let me in.'

'You can't,' I grunt, sulking, when I can't think of anything better to say. 'You have to go to work.'

'I'm billing the morning to you today,' Ulf answers, still without losing his composure, and without lighting a cigarette. He wouldn't let go.

'Fuck!' I leap up from the sofa. 'How is it possible to be so stubborn? I can't understand it. Am I just supposed to travel up there and row around this bloody island searching for a guy everyone says is dead, drowned and gone for ever?'

'There's another reason I want you to go up there.'

'And that is?'

'Elisabeth.'

'My sister? What has she got to do with this?'

'Nothing.'

'So?'

'When did you see her last?'

I shrug in defiance.

'I want you to talk to her when you're up there.'

'About what?'

'About yourself, and what you've been through.'

'Why?'

'Look on it as a necessary part of your new life, Thorkild. You're no longer in charge of interrogations; the information gatherer is dead, stripped of his reputation as well as his job title. Now you're just like all the rest of us, an ... information sharer. No matter how hard it may be to accept that.'

Ulf's lack of subtlety in laying down the law can be strong meat for people with delicate egos and screwed-up self-confidence. Luckily for me, my ego is dead, and my self-confidence has moved out and found itself better rates elsewhere.

'You need responsible people around you,' Ulf ploughs on. 'Support and sound frames of reference from outside of your therapeutic circle, not just within it. And the first thing I'd like us to include in this paradigm is your sister Liz, who I know you're fond of, more perhaps than you're willing to admit to yourself. What's more, I've mulled over what we were talking about yesterday, and I'll write a prescription for OxyNorm for your trip,' he continues. 'Then you'll have something that will work quickly if you ever need it. What do you think?'

My mouth starts to water and I can feel something start to tingle in my belly and rise along my spine to the back of my head. 'How much?'

'The same as last time.'

'What if I have to stay there for more than a week?'

'Then I'll email you a new prescription wherever you are.'

I put the phone down on the TV table and press my knuckles against my mouth. The pain in my cheek has gone, melting away the one damned time I needed it. 'Shit, shit, shit!' I hiss into my knuckles, gnawing at them until at last I take a deep breath and pick up my phone. I lift the corner of the blanket at the window once more. 'OK,' I whisper into the receiver. 'I'll go. Come on up.'

Ulf is still on the phone when I open the front door. He nods and pushes his way in, before tearing the fleece blanket off the window and flopping down on the sofa, which creaks under his weight. 'OK, then, and when does that leave for Tromsø? Half-past three? OK.' He clicks his fingers and nods in the direction of the kitchenette.

I shrug. 'What?'

'Ashtray, for Christ's sake!' Wrenching off his bag, Ulf fishes out his wallet and a packet of Marlboro Gold, pulls out his credit card and produces a cigarette that he lights with an angry gesture before inhaling greedily: 'One way, yes.'

The acrid cigarette smoke forces its way in through my nostrils and settles under my cheek, just below the skin. I turn away and step into the bathroom, where I take out my washbag and put my razor, medication pack and bag of medicines down beside a toothbrush and the rest of my toiletries. Afterwards I head out to the kitchen to pack the coffee machine and portable radio, as Ulf rounds off his phone conversation.

'Hi there,' Ulf yells out of the cloud of smoke on the sofa when he catches sight of the coffee machine, which I'm about to wrap in a towel. 'You don't need that. They've got coffee in the north too, you know.'

'I prefer my own,' I protest.

'But for God's ... no, never mind,' he waves at me and pulls a grimace. 'Yes, go ahead, just take it, pack the kitchen sink for all I care.' He then goes back to his phone conversation with a start. 'Yes, hello? It comes to how much, did you say?'

Once he has concluded his conversation and lit another cigarette, he turns to face me, nodding as he holds the smoke down in his lungs: 'Do you know what?' he says when he blows the smoke out at last. 'I think, in fact, this trip is going to do you good. Real good ...'

CHAPTER 7

It's already dark outside when the plane lands at Tromsø airport. A fine layer of fresh snow blankets the ground. I collect my bag and head out into the cold air to find a taxi. Fifteen minutes later, my sister is staring at me in disbelief.

'Thorkild?!' She folds me in her arms.

'Hi, Liz.' It feels good to hold her and I don't want to let go when she makes to pull away.

'Is everything OK?' She runs a plump finger over my cheek, while her round eyes study me intently.

'Top notch,' I answer.

'When did you get out?'

'A couple of days ago.'

'What are you doing here?'

'An inquiry,' I answer.

'Inquiry? Are you back in the police?'

'No.'

'B ... but,' she blurts in confusion.

'Can't I come in?'

'Yes, of course.' Liz ushers me into the hallway, where we stand in silence looking at each other. She looks worn out. She'll be fifty this summer, but looks older. Her eyes are swollen, as if she's been crying recently. Her hands are rough and pudgy, and she's still very overweight. The years since we last met have left their marks on both of us.

'You look sad, Thorkild.'

'Don't give it a thought. How are you doing?'

She withdraws ever so slightly. 'I'm absolutely fine.'

'Is he still hitting you?'

'Thorkild, promise me you won't …' I see the desperation mount in her eyes at the same time as I feel myself seethe inside.

'I'm only asking if your husband is still knocking you about. Judging by the bruises on your neck and arm, it looks as if he's taken his hobby to new heights.'

'I can't face this now. Arvid and I … things are going well at the moment, and you can't come here and spoil it all. I don't want … I won't allow it.'

I shake my head and step inside the living room, still wearing my shoes.

'Where is he?'

'Thorkild!' she pleads in that trembling, hysterical voice she has acquired after all the years of living with a violent long-distance lorry driver who can't manage to keep his hands off her, in the worst way.

I can hear a creak from above and take three or four strides to climb the stairs before throwing open the bedroom door.

Arvid is sitting up in bed, shifty eyes skulking behind tufts of dark, greasy hair.

'What the hell are you doing here?' he asks just before I cross the room and punch him. Arvid falls backwards, rolls out of bed and lies there with his head hidden beneath the bedside table.

The next moment, Liz comes wheezing into the room and starts to tug and yank at my jacket, crying and yelling all the while.

'What have you done? What have you done!'

Arvid finally struggles to his feet, using his hand to cover his ear. He glares at Liz. 'Do you see? See what you've done. That bastard there is dangerous; I've always said that, haven't I? He's a monster, do you understand that?'

Spitting blood, he wipes his hands on his singlet.

Liz tears herself away from me and runs over to her husband. Her fingers caress his face as she whispers soothing words.

Arvid pushes her hand away and lunges towards me. 'I think you should be damn careful, or else I'll report you, and then they'll take you right back inside, you know that, you fucking murderer!' he barks at me as he passes. 'And if you're here when I get back, you'll have to take the blame for what happens.'

He digs his elbow into me on his way out and slams the door behind him.

———

'Is this honestly what you want from life, Liz?' She has put out coffee, cakes and biscuits, and we're sitting together on the sofa in the living room, unchanged since the last time I was here. The only new addition is a black leather chair in front of the TV set. Arvid's, of course.

'It's not how you think.' She looks at me and decides to turn the conversation in a different direction. 'Have you spoken to Mum since you got out? She asks after you whenever I phone.'

'I haven't had time yet.'

'They say she's deteriorated lately.' Liz looks down at the cake dish. 'If only it wasn't so expensive to fly to Oslo. And now that Arvid's on disability—'

'And Dad?'

'He's still active. I saw him on the news a while back in connection with the building of a new aluminium plant in Iceland. They said he was the leader of a new environmental protection group. Some sort of guerrilla group. *Kæfa Ísland*, they call themselves, apparently.'

'Choke Iceland?' I laugh as I picture the smouldering eyes and long, shining, silver hair of the man who screams in visceral euphoria every time the police attempt to drag him and his compadres away from yet another industrial development, yet another protest march against the ruling powers of capitalism on our volcanic island. 'Nothing changes.'

'You remind me of him.' Liz's gaze sweeps across the scar formations on my cheek before she looks me straight in the eye again. 'It's almost like seeing him, the way I remember him from when we were little.' She lets out a sigh that causes the whole of her massive body to wobble on the sofa.

'Apart from the hair, of course. Why do you always cut it so short?'

'In what way?' I ask expressionlessly, watching her outburst of joy wither and die. 'In what way are we alike? It must be that totally unique chemical link of ours you're talking about. This corrosive fluid we secrete that smothers and destroys everything and everyone we come into contact with. Is that what you see?'

'Thorkild, that wasn't what I meant, you know that. I know you never wanted ... that what happened to ... that you never ...'

'So what is it, then?'

'I just ...' she says, grabbing another biscuit. Her eyes find their way back to my cheek and its lacerations. 'You who were always so good-looking,' she wails, burying her face in her hands.

'Come on, Liz,' I say, putting a hand on her arm as I make an effort at a painless smile. 'We can't all be as attractive as you when approaching fifty.'

'My God, give it a rest, Thorkild,' she gasps, looking at me through her fingers. 'Stop teasing, please.'

'What?' I spread out my arms. 'I mean it.'

At last she takes her hands from her face.

'But Thorkild,' she ventures once she has finished eating and wiped her hands on her trouser leg. 'I don't think you can stay here.'

'Take it easy, Liz, I'm not planning to stay,' I say. 'I no longer have a driving licence and I need help to get myself a hire car.'

'It's not so easy for Arvid either.' Her gaze lingers on the empty cake dish, as if she's seeking strength among the crumbs, to press on with these lies she tells herself each day so as not to go under. No matter how many times I kick that lazy, vicious brute that she stays married to, he will continue to lash out, and she will always take to the cake dish in a quest to find strength to go on. Liz still believes it is all just a phase they have to go through, and that if she stops doing all the things that make him hit her, then it will all work out.

CHAPTER 8

I leave the rental car Liz has arranged for me in the hotel's car park and make my way to reception, where I check in and ask for a parking permit. It is a three-hour drive, including two ferry trips, between Tromsø and Blekøyvær, where Arne Villmyr has organised a meeting for me with the local police chief next morning.

I have been allocated a hotel room with some sort of view: square buildings, streetlights and stretches of tarmac. The Paris of the North is hiding somewhere in the darkness. I draw the curtains, open my bag and unpack my coffee machine, even though there is an electric kettle in the hotel room.

I'm late. It is after half-past seven, and my body is aching. The craving to pacify the restless spasms makes my fingers tremble as I take out my medicine pack and open the compartment containing my evening dose.

The tablets look like tiny insect eggs as they roll around in my hand. I pick out the two orange Risperdal antipsychotic tablets, return them to the pack, then gulp down the rest in a single mouthful. Afterwards I take out a filter and a bag of coffee, fetch water from the basin in the bathroom and start up the coffee machine.

As soon as the first drops trickle down into the glass jug, I switch on the travel radio and turn off the lights, one by one.

I undress and creep beneath the quilt. My body has already begun to tense. A murky gloom settles and takes root deep inside me, opening doors I can't manage to open by myself. 'At last,' I sob as I press my torso harder against my knees. 'At last, I'm ready.'

I lie there waiting, but nothing happens. The air-conditioning hums, the cold polar light forces its way through the curtains, and I remain still.

In the end I sit up again and take out the pack of fast-acting OxyNorm. I press out two of the eggs, toss them down my throat and lie down again.

After another lengthy spell of more waiting, more pills and paroxysms of despair, I finally get dressed and go out.

On the other side of the bridge I spot the Arctic Cathedral outlined against the polar sky, dark and cold. My stroll skirts the edge of the harbour until I end up at a shopping centre beside the Coastal Steamer quay.

Inside the centre I cross to a shop that sells perfume. With care, I take down bottles from the display shelf and sniff at each and every one. After checking an assortment, I select a transparent glass bottle with black, oily contents and a silver lid, and take it to the checkout.

'Shall I wrap it for you?' asks the shop assistant, a woman in her fifties with immaculate makeup and black-tinted hair, thin lips painted red and dark eyes.

I nod distractedly.

'She's going to love this,' she says, smiling, as she hands me the carrier bag with the wrapped perfume bottle.

'Yes.' I peer down into the bag at the scarlet wrapping paper. 'Maybe you shouldn't have wrapped it,' I begin to say as the woman clears her throat and an elderly woman in a quilted jacket comes up beside me, carrying a perfume bottle with a bee on the lid and the word *honey* written in fine black letters on the side.

'Oh well, then.' The shop assistant blinks. 'You can always take off the wrapping paper before you give it to her.' She blinks twice and turns to the woman with the honeypot. 'You'll love this,' she says with a smile. 'Shall I wrap it for you?'

I close the bag and leave.

As soon as I'm back in my room, I take out the carrier bag and place the wrapped perfume bottle on the bed beside the pillow. I shed my clothes and stretch out on my back beside the headboard, peel off the tape and tear away the gift-wrap.

The fragrance is escaping from the box even before I've opened it. My eyelids feel heavy, and the tingling in my legs is slowly subsiding. I have to hurry, and take the bottle from the box with trembling fingers as I struggle to contain my excitement.

The silver top slips off easily, and I tug at the catch that holds the pressure mechanism closed. A spray of scent particles shoots out of the nozzle and hits me in the face the moment I press the top. I sneeze and spray some more, before sliding down into the bed and closing my eyes.

I sprawl face-down on the pillow fabric while I wait. After a while, I open my eyes and sit up. The air conditioning is sucking the fragrance from the room and filling it with more hotel air, sterile and cold.

I haul myself out of bed and check that the windows are properly closed before I crawl back under the quilt again and spray my face once more. This time I also spray the perfume on my hands and hair, before creeping back under the quilt.

'Fuck it!' I stand up and grab the perfume bottle, twist off the spray mechanism and put the opening to my mouth. The particles of scent wash over my tongue and down my throat. Dropping the bottle, I fall sobbing on the bed, dragging the quilt over me.

'Why won't you come?' I whimper, burying my face in the sheet as my body contracts with convulsions. 'Can't you understand that I need you?

CHAPTER 9

My first day with Frei, Stavanger, 22 October 2011

Outside the villa in Storhaug, the water in the fjord at Hillevågsvatn sparkled in the warm sunshine. Tiny specks of dust danced in the rays spilling through the massive west-facing windows. I had almost finished a rehearsal of the formal complaint with Frei's uncle, Arne Villmyr, a business lawyer in the oil company that had lodged the complaint. We sat at either end of a glass coffee table, while Frei curled up in an armchair at the window reading a book and wearing earplugs.

'She's working on an assignment.' Arne Villmyr was in his mid-fifties. He had a subtle tan and healthy glow, and though his hair was thin, it was dark and neatly combed, plastered to his crown.

'Oh.' I turned to Frei in the brilliant sunshine. 'What are you studying?'

Frei did not answer – she didn't even glance up from her book.

'Law,' Arne Villmyr volunteered for her. 'At university here in Stavanger.' He rubbed the point of his smooth-shaven chin before making a sign to his niece.

'Frei!'

'What is it?' Frei switched off the music and sat up straight in her chair.

'Didn't that assignment of yours involve something about the cases of police violence in Bergen in the seventies?'

'Why do you ask?'

'Our friend here from Internal Affairs probably knows a thing or two about the subject.'

'Er ...' I cleared my throat and flipped down the screen of my laptop. 'I don't know how much help I can be. Of course, the cases in Bergen were one of the reasons for changes being made in procedures to do with investigating police officers and prosecutors at the end of the eighties, and for the previous organisation, SEFO, being set up to investigate internal issues. But I don't have any professional expertise beyond what you can read for yourself in books and dissertations about the subject, unfortunately.'

Arne nodded absentmindedly as Frei looked at me through narrowed eyes. 'Do you know what?' She twisted the earplug cords between her fingers. 'Maybe you can help me all the same.'

'Oh?' I turned to her again.

'Perhaps I could interview you.'

'About what?'

'I don't quite know. But maybe you're more interesting than you first appeared. Maybe I'll identify a different angle for my assignment, or ...' She hesitated slightly before breaking into a broad smile: 'Maybe we're the sort of people who'll end up falling in love with each other?'

'Good God, Frei!' Arne opened out his arms in consternation. He was about to say something further when Frei burst out laughing and turned her face away, switching on her music and opening her book.

'I'm sorry.' Arne tasted his coffee, smacking his lips. 'Well, shall we see if we can wrap this up soon? I'm sure I'm not the only one with more to do today.'

We completed the witness statement and went through the complaint that Arne Villmyr's oil company had submitted to the deputy judge in the compensation case they had already lost in the District Court. As early as this, I knew that the complaint was going to be shelved. Outside, the sky was clouding over. The flat calm water had taken on a film of broken ripples. I had just packed away my laptop and said thank you for the coffee, and was standing up to leave when Frei pulled out one of her earplugs and looked at me.

'Well? What do you say?'

'To what?'

'You and me at Café Sting, tomorrow at six o'clock?'

'I ... I ...' I was going to say something further, but Frei had already returned the earplug to her ear and her eyes to her book. Arne Villmyr had disappeared into the kitchen.

I could have said so even then. Told her that I knew, that I'd known all along – but I didn't. Without a word, I simply turned on my heel and left.

FRIDAY

CHAPTER 10

Outside the hotel room I can hear the traffic come to life in the city of Tromsø. My throat is sore and my tongue feels rough and dry. Her scent is gone, leaving only the taste of essential oils and solvents seared into my mouth.

I switch off the alarm clock, get out of bed and make for the coffee machine to pour a cup of the coffee I brewed the previous night. I've received a message from Anniken Moritzen asking me to call her when I reach the lighthouse.

An hour later I'm on my way across the Tromsø Bridge in the hire car, heading north. You can see the newly fallen snow edging down the highest mountain peaks, against a backdrop of grey sky and a foreground overlaid with shrivelled leaves and yellow grass.

A few hours and two ferry trips later and I've arrived at the main hub of the district, Blekøyvær, which comprises various official buildings, two grocery stores, a garage, a craft shop selling knitting wool with a tanning booth in the basement, and a roundabout.

The local police station occupies part of a square two-storeyed building: green with white window-frames. The influence of Russian post-war architecture is staggering. I learn from the woman in reception that they share it with Health and Social Services on the upper floor. The police station is at ground level.

'Bendiks Johann Bjørkang,' says the local police chief, a man in his sixties who speaks with a thick local accent and sports cropped brown hair and a luxuriant moustache. A strong hand grabs mine. 'You're Thorkild Aske?'

'So they say.'

'Who?'

'The people who say that kind of thing.'

My attempt at humour leaves him cold, so after running a finger over his chin, the police chief invites me into his office.

'A kind of ... private investigator?' He sits down in a chair behind the desk and rests his hands on his belly.

'No,' I answer, taking the rickety wooden chair, with a folded newspaper shoved beneath one leg, that stands in front of his pale pine desk.

'What are you doing here, then?' He drums his fingers on his paunch.

'I'd prefer to call it a one-off, forced on me by personal circumstances and accepted by yours truly owing to a lack of resistance and acute financial need.'

'I see.' Bendiks Johann Bjørkang gives a deep sigh. 'So you're here to look for the Danish guy?'

'Correct.'

'On behalf of his parents?'

I nod.

'Well, we're here to help.' He raises his eyebrows sceptically. 'I don't think you'll find all that much.'

'No, nor do I.'

'The Dane is the only case we have open at the moment. We had a Russian trawler that sank farther north in a storm a month or so ago, but everyone got out safely and apart from that it's been as quiet as the grave.' He folds his hands back over his gut. 'Quiet as the grave.'

A young sergeant in his mid-twenties, also with a moustache, comes in and stares in turn between me and the police chief with an inane expression on his face, as if he has just interrupted his boss in a moment of intimacy and is wondering if he should leave or ask if he can join in.

'Eh' – the police chief clears his throat and gestures between the sergeant gaping in the doorway and me – 'Arnt Eriksen, this is Thorkild Aske. He's here because of the Danish guy. Arnt abandoned the Tromsø hurly-burly and moved up here nearly a year ago with his girlfriend. He is going to take over the local station after New Year when I retire.'

'Hi.' The sergeant wipes his hands on his trousers before offering me a sweaty hand. He gives me a crooked smile in an effort to show his awareness of the burden involved in the dangerous position assigned to him by life. 'Arnt here.'

'Yes, and I'm here,' I riposte, shaking his hand until he takes on an embarrassed look. 'Well,' I continue, without letting go his hand, 'what did the Tromsø hurly-burly involve, then?'

Arnt looks at me and at our hands, expecting me to complete the social etiquette and let his go.

'The city is probably struggling with the same challenges as other cities,' he ventures, before clearing his throat. 'Theft of property and a steady increase in serious drugs crime. We've also seen a marked escalation in prostitution in recent years, but—'

'According to rumour, you used to work in Internal Affairs?' Bjørkang interrupts in an effort to resume control of the pleasantries. He emphasises *used to* and nods at his sergeant, who tries to sit down while we're still holding hands.

'Rumour?' I ask, letting go at last.

'Oh, you know, even though your case escaped media attention, it still ran round the police force like wildfire. It's not every day that a chief inspector in Internal Affairs is caught red-handed. It caused a furore in certain circles and a round of applause in others, from what I've heard.'

'No doubt,' I respond.

'Was there not some business later when one of our lot, someone you'd had put behind bars, tried to get his case taken up again through the legal system because of what had happened?'

'Did that help?' I ask tartly.

Bjørkang shakes his head without taking his eyes off me. 'Did you know that, Arnt?' The local police chief turns to his sergeant, who has assumed his bewildered, speechless expression again. His head moves back and forth between his boss and me, depending on who is speaking. 'That these guys here are specially trained to break down police officers who make an arse of themselves while on duty? So just look out.'

Arnt glances at me with a confused expression as Bjørkang bursts into raucous laughter that soon breaks off.

'We used to call it interviews,' I say.

'Yes, of course,' Bjørkang answers as soon as the phoney laughter peters out. 'Tell me: who interviewed you when you were arrested?'

'They brought in an instructor from the Police College,' I answer. 'With further education in the UK and an expert in phased police interviews, interview tactics, ethics and communication, psychological influences and a whole lot of memory-jogging techniques. Nice guy.'

'Arrested?' Sergeant Eriksen asks, looking as if he has just regained the power of speech.

'Oh, didn't you know?' Bjørkang winks at me. 'What is it you actually learn at Police College these days? Thorkild Aske has just been released after serving three years in jail in Stavanger.'

'You mustn't forget the time I spent in a psychiatric clinic,' I say, without taking my eyes off the sergeant.

'What had you done?' Arnt asks, clearly disappointed.

'He killed a girl in a car accident one night after work,' Bjørkang interjects helpfully. 'While spaced out on drugs.'

Sergeant Eriksen continues to stare at me, but there is now something different in his eyes. Something I recognise. Disgust at seeing one of his own who has crossed the line to the other side.

'Gamma-hydroxybutyrate, or GHB,' I answer, thinking about what Ulf had said in the car the day I was released from prison, that now it was my turn to sit at that end of the table. To be the one who adapts to conditions laid down by others, rather than dictating them. My 'pilgrimage' was what he called it. It simply hadn't dawned on me how much it was going to cost.

'I assume that was where you acquired that?' The police chief indicates the scars on my face, stretching out like a spider's web of fine threads from my eyes down my cheekbone to my mouth.

'I was lucky. My head hit the steering wheel.'

'Well,' Bjørkang begins, now in a gentler tone of voice. 'What's happened has happened. The rest will have to be between you and him up there.'

'The Head of Social Services?'

He smiles wryly. 'We're not here to dole out blame, but I'm a man who likes to know who I'm dealing with.' He nods at his sergeant, Arnt, as if to tell him that this, what he is witness to now, is what is called top-level people-management. 'I'm also of the persuasion that a man who has served his punishment is a man with a clean sheet. If I didn't believe that, I would have nothing to do with you.'

Bjørkang rises from his chair and sends his sergeant out with an authoritative gesture. 'OK, if the courtesies are over.' He makes a sign for us to go. 'Let's go out to Skjellvik and take a look at the Danish guy's boat then, before the day runs away from us entirely.'

We leave the local police station and get into a police vehicle. I leave the hire car behind. It's raining: the frozen ground is glistening where raindrops strike and the wind gusts through leaves that still cling to branches in the chill October air. Blue-black clouds are drifting in from sea.

'The dark time,' Sergeant Eriksen says, glancing at me in the rear-view mirror. 'Not everyone copes with it too well.'

CHAPTER 11

The journey to the island runs through a steep-sided valley before the bumpy road returns us to the sea. On the way we pass the occasional postwar house clad in green, white or yellow fibre-cement weatherboards and corrugated iron roofs. Some still have lights shining in the windows, and there are strips of furrowed fields, but most of them were abandoned and surrendered long ago to wind and weather.

Eventually the tarmac runs out too in favour of gravel, until we crest a hilltop that offers a panoramic view over a wide bay with scattered settlements.

'Do you dive?' The sergeant's face appears in the rear-view mirror.

'What?'

'I asked if you dive.'

Bjørkang shakes his head angrily.

'What is it?' the sergeant asks, his voice fracturing with dismay.

'The man's just out of jail, for God's sake. Where do you think he's been diving?' Bjørkang takes off his police cap and slicks back some wisps of hair. 'What sort of police work is that, eh?'

'At Haakonsvern,' I say after a pause. 'And two or three times since then.'

The sergeant's eyes have immediately returned to the rear-view mirror. 'Well,' he says eagerly, 'what did you think?'

'Hated it,' I answer, and notice the fire in the young sergeant's eyes fade out for the second time in less than a minute.

Right at the summit of the hill, there is a dark brown building and a car park, as well as a smaller, oblong building divided into three apartments.

'Skjellviktun Residential Centre,' Bjørkang says as we drive past. 'It's been there since the war. It was a barracks for the Krauts in this area.'

'Krauts?'

'The Germans,' Arnt throws in, glancing back at me in the mirror again. 'Did your family fight in the war?'

'Not at all,' I answer, scanning the way the sea inter-weaves with the landscape. On an islet farther out I spot a white building and an eight-sided lighthouse tower on a promontory close by.

'Rasmus Moritzen's lighthouse.' Bjørkang points to a cluster of boathouses at the foot of the bay. 'His boat is in one of those boathouses there.'

The wind cuts through our clothes, and beyond us the waves slap lazily together, twisting in billows of white. Fat trusses of seaweed stretch green tentacles across the pebbles on the shore. We trudge down to the boathouse with hands clamped tightly on our jacket collars, heads bowed against the rain. It's difficult to walk on the slippery stones, and I feel the muscles in my legs and up towards the small of my back flexing with the strain.

'The boat drifted in to land on Tuesday morning. We assume he was diving out beside Øyet at the weekend. His mother told us when she was up here that she'd spoken to him on Friday, but he hadn't answered when she called again on Sunday afternoon.' Bjørkang shakes his head as we approach the boathouse, where a man slightly younger than me is waiting in the doorway.

'You took your time,' the man says in an American accent, blowing on his hands. He is wearing a brown leather jacket with a wool collar of the type favoured by British pilots in the Second World War. He has a knitted hat on his head with the words *No Bullying* on it, and black leather gloves tucked under his arm. 'I suppose it was Arnt who did the driving?'

'Harvey, this is Thorkild Aske. He's here on behalf of the missing Danish guy's parents.' Bjørkang turns to me. 'And this is Harvey Nielsen. We need to see the Dane's boat in his boathouse.'

Harvey Nielsen holds out his hand. He is tall and dark, with dimples that contract into snow crystals when he smiles. 'Have you told him about the lighthouse?'

'No,' Arnt looks sanctimoniously at Harvey, towering a whole head above him. 'We haven't told him anything.'

Harvey Nielsen puts a powerful hand around my neck and turns me to the right as he points at the rocky slopes out beside the lighthouse. 'Hey! You'll get the entire story from me, since you ask so nicely.' Revealing his teeth, he tightens his grip around my neck.

'Go ahead.'

'It seems that some rich folk from the south came up here in the eighties to transform the whole little island into a conference centre for yuppies. They converted the old keeper's quarters and added a restaurant and bar, their own computer suite and gymnasium. Yes, and even an enormous disco in the basement.'

'Then the money ran out only a year after the business was launched,' Bjørkang interjects, while Arnt fiddles with his moustache, as if to assure himself that the wind hasn't blown it off. 'And the owners set fire to the main building when they didn't succeed in selling it, in a botched insurance scam.'

'Since then, it has lain empty.' Harvey takes up the thread. 'Until the Danish guy arrived and began to renovate it last summer.'

'An activity hotel,' the local police chief snorts. 'What on earth is that?'

'He was capable. I was out there and had a look. He'd taken an old Nordland boat and converted it into a table, and the bar ...' Harvey gives a wide grin. 'The bar is absolutely fabulous.'

'I'd never have gone out there on my own,' Arnt mumbles as he gazes out at the lighthouse.

'Well, we'll have to take a look at that boat before it gets dark.' Bjørkang slaps Sergeant Eriksen on the back and steps inside the boathouse with Harvey and me trailing behind him.

The wind and rain tear and buffet the corrugated boat-house roof. Bjørkang and Harvey strip off the tarpaulin and place it on the earth floor against one of the walls.

'So here you have the boat,' the local police chief says, stroking it with one hand.

The boat, about six metres long, is a blue and white RIB – a rigid inflatable boat – with a 150-horsepower Evinrude engine and a fibreglass hull boasting the trademark Zodiac

Pro. I clamber aboard. There is a rope and an air cylinder inside the boat, as well as a wooden chest filled with what seems to be old bits and pieces collected from the seabed.

'Quite a craft, eh?' Harvey picks up a square box covered in dry strands of algae and tiny white shells and passes it to me.

'The Danish guy obviously collected old rubbish,' Bjørkang mutters, nodding at the old transistor radio I'm holding in my hand. It had once been white, with a blue border, and I can just make out the number three together with some letters, P b K A, up in the left-hand corner of the shell- and algae-infested apparatus.

'No GPS?' I ask, nodding at the empty casing beside the steering wheel.

'Maybe it fell out while the RIB was drifting,' Bjørkang answers casually.

'Why didn't this fall out, then?' I ask, lifting the transistor radio. 'I assume this was lying in the boat and the GPS was attached to the centre console?'

'Who knows?' Bjørkang shrugs, heaving a sigh.

'OK.' I replace the radio in the RIB. 'You said he was most likely diving in a place called Øyet. Where's that?'

'Øyet is an underwater mountain right between the lighthouse and the islands on the other side of the fjord.' Bjørkang looks at his watch and heads to the boathouse door to point over to where a black pole with a topmost flashing light stands on a rock that juts out of the sea. 'Put there at one time to help boats that venture into the reefs, through the Grøt Sound.'

'It's called Øyet – the eye,' Harvey explains, 'because when the sea is calm, it looks as if someone is looking up from the depths at the sky. Over the years there's been many a ship and crew that has gone down after steering too close to that reef.'

'I presume that's why he was diving? Because of all the wrecks?'

All three nod in unison. 'We had divers down there but they didn't find anything,' the local police chief begins after a brief silence. 'We'll just have to wait,' he goes on, 'let nature take its course, and he'll come up again in time, you'll see.'

'When the crabs are finished with him,' Harvey adds.

It is growing dark, and the sky is disappearing behind black clouds.

'I'll need to take a trip out there,' I say. 'To the lighthouse.'

Bjørkang looks at his watch again. 'We should wrap this nonsense up now,' he says, before spitting on the ground. 'There's more bad weather forecast all weekend. You go home to Stavanger, Aske. I said the same thing to the Danish guy's mother when she was up here yelling and shrieking to be taken out there in the middle of a gale, even though it would risk the crew's lives. We'll call you when he pops up again. They always surface, these bodies in the sea. But it can take time.'

'His parents want me to go out there, so I'm going to do that, unless you have legal grounds to prevent me.'

'OK, OK.' Bjørkang flings out his arms in annoyance. 'If you want to stay up here waiting, that's fine by me. We can't deny you that. But be a bit cautious, that's all I ask. You're no longer a policeman.'

'I'm just here to help,' I answer. 'The same as you.'

'Great,' Bjørkang grunts. 'Remember, if there's anything else you think we should waste our time on, then it'll have to wait till Monday. OK?'

I nod and Bjørkang consults his watch one last time before beckoning Arnt and indicating they are about to leave.

'Cognac and card games,' Harvey chuckles once the two policemen are sitting in their car again. 'At weekends he drinks cognac and plays cards with the other old boys out here on the islands. Public servants, you know.' Harvey laughs. 'Crime is something that only takes place on weekdays between the hours of eight and four up here. Didn't you know that?'

'I see,' I say, sighing, when it dawns on me that I've left my car parked outside the local police station.

'Why do you actually want to go out to the lighthouse?' Harvey asks after closing the boathouse door and attaching the padlock. 'I was there with Bjørkang after we found the boat. There's nothing out there.'

'His mother wanted me to. As for myself, I just want to get it over and done with. I'm freezing to death.'

'OK.' Harvey beats a tattoo with his fingers on the boathouse door. 'Well, you can come with me tomorrow morning if you like. I'm going past there on my way to the farm.'

'The farm?'

'Mussels.' He smiles again. 'That's where the money is.'

CHAPTER 12

I get a lift from Harvey in his pickup. He has come up with the idea of having me stay overnight, since we'll be going out to the lighthouse next morning, so as to spare me the trouble of commuting the hundred kilometres and two ferry trips from my hotel room in Tromsø. His house is located in a new housing estate with a view over the fjord and the lighthouse on its little island.

We go inside and sit at the kitchen table. Harvey brings two mugs of coffee. In the background, his wife is fussing after a boy of about six; she's in a hurry, trying to get him dressed.

'We're leaving now,' she says, taking the boy with her to the hallway, where she struggles to put on her jacket, using only one hand.

'Come here.' When she approaches us, Harvey pulls her down on to his knee and kisses her hair. 'This is Thorkild Aske, a former policeman, here to search for the missing man.' He kisses her again. 'And this is my darling wife, Merethe.'

Merethe extricates herself from Harvey's embrace and offers me her hand.

'Hi,' I say, straining to smile with both corners of my mouth at the same time.

'Hi, Thorkild,' she says before abruptly releasing my hand and heading off to the hallway, where the little boy is now flinging shoes around, making sounds of gunfire to accompany the missiles.

'Merethe works as an occupational therapist up at the residential centre. Senior yoga, healing, crystal therapy and that sort of thing. She'll soon be a real celebrity into the bargain, won't you, sweetheart?'

'What?' Merethe yells from the hallway.

'You! You'll soon be a celebrity,' Harvey answers. 'Come here, and you can have a go with Thorkild. I'm sure you'll find some ghosts hanging around him too.'

Stumbling to her feet with her hands full of shoes, Merethe lobs one at Harvey in the kitchen. 'Not now, Harvey. Can't you see I'm running late?'

Harvey turns to face me again as his wife rushes out of the door with the boy in tow. 'She's always in spiritual unbalance at this time in her cycle.' He laughs loudly, slapping himself on the knees.

'What's she famous for?'

'Clairvoyant,' Harvey replies. 'She's been hired for the next series of *Spiritual Powers*, if you follow that programme. They're starting filming now and into the New Year. She's been booked for four episodes in the first instance. Big shot, eh?'

'I don't watch as much TV as I should,' I venture, before Harvey gets suddenly to his feet and disappears down into the basement. He emerges with a plastic container that he places on the table between us.

'You look like a guy who won't say no to a drop or two.' He unscrews the lid and dilutes the coffee with the crystal-clear spirit. 'Don't they serve laced coffee in Iceland? That's where you come from, right?'

'Yes, and yes – they're just not so generous with the coffee.'

Harvey laughs and we sit drinking in silence as we stare through the kitchen window at the sea and the polar night now falling over the landscape.

'What would life be without children?' Harvey says in the end. 'Do you have any?'

'No.'

'Married?'

'A long time ago.'

'What happened?'

'I went to the USA, and she went to Gunnar.'

'Gunnar?'

'Gunnar Ore. My former boss in Internal Affairs.'

'Bloody hell, man, sounds like a real bitch.'

I shrug. 'We weren't getting on very well.'

'So that's why you became a private investigator?'

'Something like that,' I answer wearily. I can see the lighthouse looming over the top of the small island in the murk. The whole place is grey. It won't be long until it, too, is swallowed by the blue-black wall.

'Harvey Nielsen,' I begin after another hiatus. 'Is that a northern Norwegian name?'

'Absolutely.' Harvey guffaws again. 'My forebears emigrated to Minnesota in 1913, before my great-grandfather came back during the First World War and was gassed to death in a field in northern France not long after.'

'And you? How did you find your way back to the Promised Land?'

'Signed on board a fishing boat after college and ended up by chance in Tromsø, where I met Merethe, working in a pub in the city.'

'And became a mussel farmer?'

'Among other things. Merethe's family had a small-holding where we kept sheep for a while until early in the noughties, when I happened across a course in mussel cultivation. I applied to the government's trade and rural development fund for support and got started. I began with a small patch, a few plastic containers, some old herring net, tractor parts, electric cables and concrete foundations I poured myself. I built the floating rig myself too. Most mussel farmers went under in the first few years, but we held on, gritted our teeth and kept going until we came out on the other side – changed,' Harvey says with a grin. 'And in May we delivered seven tonnes of mussels. Next year we're going after ten.'

'Where is the farm?'

'In a cove farther north, where Merethe's family had their farm before it was wound up. We went on running it for a while after her parents had moved into residential care, and kept it going for a few years, but there's no money in small farms any longer, nothing but work and hard slog.'

'Do you miss the States?'

'No,' Harvey answers. 'Not at all. I didn't belong there – I knew that when I was a kid. The sea, you know. It's in the blood, breaking against the artery walls and calling out.' Harvey knocks the bottom of his mug on the kitchen table before letting his gaze roam out of the window where the streetlights are twinkling in the ice-cold autumn darkness. 'I could never leave this place. Never.'

I'm aware of the alcohol starting to have an effect on me. My body is filled with an intense heat I haven't felt for a long time.

'You said you went to the US.' Harvey looks at me through limpid grey eyes. 'What did you do there?'

'Professional development,' I answer. 'Or was it to escape from a broken marriage? It's hard to recall now, looking back.'

Harvey raises his mug in a silent toast. 'Hear, hear!' Then he drinks and puts down the mug, all the while looking at me, with a half-smile playing on his lips. 'So, why did you go there, in truth?' he eventually asks.

'Here in Norway the police force follows a standardised interview technique called KREATIV,' I begin to explain.

'And that is?'

'KREATIV focuses on the accused's witness statement. The aim of the interview is not necessarily for the accused to confess, but rather to reduce his chances of putting forward plausible cover stories and counter-strategies along the way. In Miami there was a totally unique opportunity to learn about more advanced interview techniques and investigative psychology from no less a person than Dr Titus Ohlenborg.'

'And he is?'

'Heard of KUBARK?'

'Nope.'

'KUBARK was a total of seven interview manuals intended for training specialist staff in the CIA, the army and other special forces. The first one was published in 1963 during the Cold War. One of them was purely a training tool for lead interviewers specially linked to counter-espionage, and came with a series of techniques meant to be used to facilitate breaking everyone from defectors to refugees, agitators, agents and double agents, and discover whether they were bona fide or not. This manual was written by the good doctor himself.'

'Spy shit? Really?' Harvey rolls his eyes and gives a crooked grin. 'You don't strike me as the type.'

I shrug. 'Ohlenborg is a professional psychologist and began his career by studying the interaction between people and buildings, before he transferred to the CIA. Now he teaches everywhere from official investigative agencies to private security firms such as Blackwater, DynCorp and Triple Canopy.'

'And Norwegian policemen?'

Nodding, I signal to ask for a refill. 'The problem with manuals like KUBARK and more recent European methods like our own KREATIV is always the same.'

'And that is?' Grabbing the plastic container, Harvey stands up and leans across the table to pour out more.

'How do you interview someone who knows and can do exactly the same as you? Someone who may have received and mastered the same training as you?'

He puts the container on the floor and flops down into the chair again. 'I see,' he says, nodding decisively. 'How to break one of your own.'

'Correct. The special thing about Ohlenborg is that, for a number of years, he has travelled around American prisons and interviewed police officers from both local and federal police forces, who have all served sentences for a range of crimes from robbery and narcotics smuggling to hired killing, rape and serial murder.'

'Cops gone bad.' Harvey chuckles into his mug. 'What a world.'

'Intelligence organisations, the military and police all face the same challenges in an interview situation when interrogating one of their own. People who have themselves conducted hundreds, maybe thousands of interviews throughout their careers, who are familiar with the methods, who might have perfected these over the years with that very idea in mind – the day when they're caught and everything's at stake.'

'So what do you do to break them?'

'Your own experiences, training and belief in your own capabilities count the highest in every interview situation. But what you eventually learn is that these very men, no matter how good they are and how accomplished at the game, and whatever their portfolio of life experience might contain, they can't succeed in disguising the human being, and their humanity is the way in. The bottom line.'

'You've lost me, man.' Harvey shakes his head.

'We are all ruled by the elementary strings in our emotional register. The difference is what happens to each and every one of us when someone plays on these strings. No matter what,' I plough on as I rotate the mug between my hands and stare down at the muddy liquid. 'After nine months of travelling, Dr Ohlenborg fell ill and had to go through a new type of radical radiation treatment for a brain tumour, and I came back to Bergen and Internal Affairs once again.'

'So why did you give up the police?'

'Another time,' I whisper. 'Another time altogether.'

Outside, the rain has frozen into hail, tapping on the kitchen window before falling back into the darkness again.

'You said it was the Danish guy's parents who had hired you,' Harvey finally says.

I nod.

'To do what?'

'I don't quite know,' I reply. 'Search. Hope can actually be bought, you know.'

'Hope?'

'As long as they pay, I search. As long as I search … there's hope that I might find something.'

'Find what, do you think?'

'A magic key to turn back time.' I peer into my coffee mug again as Harvey fills it up. The aroma of the alcohol tickles my nose, heating, breaking open my ruined tear ducts and beckoning clouds of memory from the depths of my mind. Nodding, I open my mouth and swallow. Huge gulps.

'Do you ever find it?' Harvey asks, half mockingly, and looks at me. 'This key?'

'Never,' I reply, with a short, sharp laugh.

CHAPTER 13

My second day with Frei, Stavanger, 23 October 2011

Café Sting stood next to the Valberg Tower. The building was an old timber house containing the smart rustic restaurant that Stavanger and its inhabitants seemed to like so well. The woman behind the bar fixed me a cup of coffee and a glass of water clinking with ice cubes that I carried over to a table at the far end of the room, where I sat down to wait for Frei.

She turned up at a quarter to seven. I was sitting at a window table peering out at the stone tower then under attack from the rain that was streaming in rivulets down the café windows.

'Awful weather,' she said, shaking off a hooded khaki-green parka with fitted pockets. She hung it over the back of the chair and shot a look at the bar, where the woman behind the counter responded by switching on the kettle and rooting around in a bowl of teabags. 'Have you been waiting long?'

'I'd have waited longer if you wanted me to.'

Tilting her head, Frei gazed at me for a moment without saying a word, before turning around and vanishing in the direction of the bar.

'Why did you come?' she asked once she had finally sat down on the cast-iron chair opposite me. She picked up three big brown sugar cubes and dropped them into the apple-green liquid in her teacup. Then she took her spoon and stirred it round lazily until the sugar had dissolved and given the contents a darker, more earthy and impure quality.

'Loneliness,' I answered. 'Without a doubt.'

'Do you think I can help you with that?'

'Almost certainly not.'

'So why?'

I shrugged. 'Because you asked.'

'My uncle wanted me to.' She placed the teaspoon under her lip and closed her mouth. 'He's expecting a visit,' she said, setting the spoon down on the dish of lemon slices.

'Oh?'

'From a man.'

'I see.'

'Robert's a couple of years older than me. Could have been my brother. Drop-dead gorgeous.' She laughed and used both hands to pick up the teaspoon. 'Uncle Arne is gay. Didn't you notice that?'

'Should I have?'

Now it was her turn to shrug. 'Uncle Arne said you were someone who read body language and revealed what we didn't want to share with others. Isn't that so?'

'No,' I said, laughing. 'Not at all.'

'So what can you do, then?'

'Conduct interviews and write reports,' I answered, using my thumb and index finger to tip the empty coffee cup over on its side. My eyes glimpsed the dregs at the bottom before drifting back up to her face.

'But you're an expert in interview techniques, aren't you? Arne said you'd just come back from the States.'

I leaned across to her and clasped my hands. 'How on earth does he know that?'

'Arne's a business lawyer for one of North America's biggest oil companies.' Frei ran her fingers through her riotous hair. 'He likes to know things about the people he meets, both privately and in connection with his work.'

'What else?'

'You're half Norwegian and half Icelandic, your father is a professional marine biologist, but is some sort of radical environmental activist at home in Iceland. Your parents divorced when you were small, and your mother moved back to Norway with you and your sister. Yes, and then you're recently divorced yourself.'

'Then you know all there is to know,' I replied, bending down to lift the bag I had stowed between my legs. I had spent half the night gathering all the documents I could find about the Bergen case and the establishment of the former SEFO. 'By the way, here's what I could find for your dissertation. The European Convention's torture committee also came out with a report that posed several critical questions about—'

'I don't need them.' Frei remained seated, gazing at me with her cup balanced in the middle of one palm, as if she were adopting a lotus position, completely unknown to me, for young, urban café-goers.

I turned my cup upside down and replaced it on the saucer. 'What are we doing here, then?'

'Getting acquainted,' Frei replied. 'In your way.'

'My way?'

'Isn't that what your job is, your expertise? Collecting facts about a person, presenting these in a controlled environment where you and your people can unearth our secrets, defects and pressure points? I just wanted us to start with you, first of all.'

'My God,' I groaned, tapping my knuckles on the under-side of my cup before chuckling to myself. 'Maybe you're right,' I said in the end, and stood up to leave. 'Or, do you know what? You're spot on.' I picked up the bag of docu-ments and gave a slight bow. 'My game, you win. Adieu.'

I stepped swiftly across the chequered tiles before making an abrupt about-turn and heading back to the table to resume my seat. 'No, as a matter of fact, let's finish the game.' I reclined in the chair. 'So, what is it you want to know, Frei? Ask away.'

'I want to know your secrets and lies,' she answered, unruffled, still with her teacup perched on the middle of her palm. Suddenly she replaced the cup and rested her finger-tips on the table in front of her. 'Before I give you mine.'

'OK. Where shall I begin?'

'Wherever you want.'

'My mother was a child psychologist until she fell ill. She now lives in sheltered housing in Asker. She has Alzheimer's; has had it for almost ten years. I haven't seen her for a long time. I don't know why.'

'Tell me about Iceland.'

'We travelled around from aluminium works to alumin-ium works, power stations and smelters when I was little. Mum, Liz and I, to watch Dad and his environmental protec-tors chain themselves to excavators and pipe couplings, cranes and dumper trucks, as they screamed with rapture and pissed themselves so that the world could see that the human race was out of control.'

'The idealist.'

I nodded absently.

'Why did you go to the States?'

'After the divorce procedure my ex-wife and I had reached a point where we ended up misunderstanding each other in every respect, every single detail, every nuance. So when I read about the course, I just went.'

'To interview criminal policemen in a foreign country?'

'Exactly.'

'Why?'

'To improve at my job, and to—'

'Understand them?'

'Yes.'

Frei suddenly gave a broad smile. 'Do you break policemen for your father, Thorkild?'

'Probably,' I answered, sounding drained. The rain was hammering on the windows. Outside, water had gathered in streams on the ground, and was running along the street on either side of the café building.

Frei suddenly burst out laughing. 'You're a cliché, Thorkild Aske,' she said, touching her head. 'Don't you see that?'

'Yes, I do. If I get myself a psychologist one day, I'll damn well say that you told me first.'

I had already decided to continue the game, regardless of the questions I would have to answer, and the confidences it would cost. The divorce from Ann-Mari and the time spent on the American south-east coast with the men and women Dr Ohlenborg and I had interviewed behind bulletproof glass had cost a great deal. For me. And sitting here with this girl in a café caused the hard outermost layer of my skin to crack. I could already feel the pulse of something new and alive in there beneath that shell. Something I had never felt before.

Frei hesitated for a moment with her smile hidden just beneath the surface, until she said in the end: 'Was that why you divorced? Had you finally understood everything about her and found nothing more to dig for? Had you charted her entire landscape, job done and on to something else, the next case?'

'Case? Such as ... you?'

'No,' Frei replied. 'You still know nothing about me. We two don't know each other.'

'You're right. Is it my turn now?'

'Your turn.'

'OK,' I said, leaning across the table. 'Tell me about yourself.'

Frei sat looking at me, only her eyes moving, gliding over every pore on my face like dazzling twin suns. 'I dance,' she said at last, and snatched up a sugar cube from the saucer.

SATURDAY

CHAPTER 14

'Hello there, what's red and says blob blob?'

'Wh … what?'

'What's red and says blob blob?'

'My God, I've no idea,' I answer, trying to wriggle away from the small creature with questions from hell.

'Ha ha! A red blob blob, of course!'

I open my eyes wide and see that I'm lying in the middle of the kitchen floor, partly hidden under the table where Harvey and I had sat sharing alcohol the night before. The same boy I had met the previous day is sitting beside me, smiling. I can see Harvey's feet over beside the cooker and am suddenly aware of the aroma of freshly brewed coffee.

'Well, then, that's the other guy up as well.' Harvey bends down to peer under the table.

'Mhmm,' I moan, struggling to my feet.

'Why are you sleeping on the floor?' the boy at my side asks.

'I don't know,' I answer, banging my head on the edge of the table as I try to stand up.

'For a private investigator, you don't hold your booze very well,' Harvey chuckles over a steaming hot cup of coffee that he places on the table above me.

'One day,' I grin ruefully, catching hold of one of the chairs to pull myself up. 'Just give me time.'

'Daddy, was he drunk yesterday?' The boy looks from me to his father.

Harvey comes over and helps me up. 'Well, something or other, anyway,' he says with a smile.

'What time is it?' I ask, burning my lips on the hot coffee.

'It'll soon be half-past five, crack of dawn,' Harvey answers. 'We're leaving in ten minutes.'

The inside of my skull is thumping, my sinuses feel filled with cement, and my cheek is throbbing like mad.

'It's foul outside today,' Harvey says over his coffee. He seems surprisingly fresh and sharp, considering the night before. 'I've laid out a set of woollen underwear for you to borrow, as well as a pair of boots and a hat.' He smiles again as he comments: 'To ward off the cold.'

'Thanks.' I shiver at the view outside the window. It's still dark outside, and only a pale morning light above the highest peaks shows that day is breaking.

'No problem, man.' Harvey raises his cup in a toast. I down a few more swigs until the grumbling in my belly and the general unease I feel before consuming my morning dose forces me to my feet and into the bathroom.

The sight that greets me in the mirror is enough to send any living thing from the underworld racing back to where it came from. I take out my morning pills and wash them down with water from the tap. Afterwards I squeeze a drop of green children's toothpaste on my forefinger and do my best to brush my teeth. The rest would have to wait. What's broken can't be mended.

En route to the hallway, I meet the young boy again.

'Hey, you.' He imitates his father by standing with one hip lower than the other, his arms crossed over his chest: 'What's red and says blob blob?'

I stare at him sternly in the hope that my withering gaze will make him appreciate that he is insane and needs to make himself scarce, but it doesn't seem to have any kind of effect. So I give up, smile and answer: 'Yes, wasn't it some sort of blob blob, red in colour?'

'Silly! It's a cranberry with an outboard motor, of course! Ha ha ha ha!'

Outside, everything is greyish-black. Even the houses on the estate seem to have lost all trace of colour. A magpie is sitting on the roof of a bird table in the garden, staring at us with its head cocked. It takes off just as Harvey activates the lock on the pickup door. He starts the engine and we drive down the winding road past the centre towards the boathouses at the foot of the cove.

Yesterday's wind has died down and the outside temperature has plummeted. The air is raw, making it hard to breathe without coughing. The pain in my cheek is pounding, piercing, a constant presence. To top it all, it is time for me to acknowledge that my digestive system is out of order, and these stomach pains I've been struggling with recently are not going to disappear of their own accord.

'I'll pick you up as soon as I've finished at the farm,' Harvey tells me.

'OK,' I answer just as I catch sight of an old man in a plastic boat on its way to the coast. As soon as the boat lands, the old guy jumps out and busies himself pulling it up the foreshore.

'Do you need help, Johannes?' Harvey goes down to the man even though he is shaking his head. He hauls out a box of fish and starts to gut them on the pebbles.

'I've got a genuine private investigator here with me,' Harvey says, leaning on the gunwale as Johannes cuts open the belly of a large cod and tears out the entrails. A couple of seagulls are roosting on the fish flakes beside the boathouses. 'He's here to find the Danish guy, you know, the one who was staying out at the lighthouse.'

Johannes steals a glance at me again before shaking his head and tossing the entrails out to sea, just as one of the gulls takes off from the drying frame and comes swooping towards us. 'The Danish bloke is dead.'

'Yes, but—' Harvey ventures, only to be interrupted by Johannes, who has flung the gutted cod back into the fish box and pulled out another.

'You know ...' Johannes presses his lips together so that it looks as though he's chewing the inside of his cheek, 'the Sea Ghost sails high on the crests of the waves at this time of year. There's bad weather in store.' Using his knife to slice open the fish's stomach, he rips out what's inside and throws it out to sea. 'Take good care of that southerner. You remember what happened to the Russian trawler in the last storm.'

'He isn't a southerner,' Harvey retorts laughingly. 'He's an Icelander.'

'I see.' Johannes lifts another box of fish out of his boat and sets it down between us on the pebbles. 'Is there a difference?' He crouches down and continues gutting as we head back to the boathouses.

'Who's the Sea Ghost?' I ask once we're skimming through the waves in Harvey's RIB on our way out to the lighthouse. The wind has whipped up over the past few minutes, and I grab at a rope for support.

'Haven't you heard of the dead man who sails around in half a boat, dressed in leather oilskins and giving warning when someone is about to die?'

'Sounds like a nice guy,' I mutter, tugging my hat further down my forehead. 'Johannes' more playful big brother, maybe?'

'The sea gives and the sea takes away. Up here old superstition is something you hug close to you when storms are raging at their worst. There are more things in heaven and earth, Thorkild Aske. Especially up here in the north. Don't you have those traditions in Iceland?'

'Yes, of course.'

He reduces the boat's speed and turns to face me. 'Do you believe in ghosts, Thorkild?'

'Ghosts?'

'Apparitions, dead souls who wander among us and all that stuff?'

'I …' I start to say, but stop. For a moment I simply sit there without moving a muscle while I feel the cold and wind pummel my face.

'I remember when I was little.' Harvey brings the boat to a complete standstill, so that it bobs up and down with the rhythm of the restless sea. 'You could hear the voice of a crying child from the forest around our cabin back home in Minnesota. This was in winter, when swamps and lakes froze fast, that we could hear it. Heart-breaking childish sobs like an echo between the tree trunks when the frosty mist settled on the forest floor. It made the hairs on the back of your neck stand up, I can tell you.'

'I see,' I mumble as my eyes roam restlessly over the cold dark surface of the ocean.

'Later, a couple of the smallest lakes were drained in connection with building another big cluster of cabins. The excavators found a child's body in one of the bogs: a corpse they thought had lain there for more than a century. After that, the forest was silent. What do you say to something like that?'

'I don't know.' I glance up at the sky where dark clouds are billowing in from the ocean. Ahead of us lie the lighthouse and other buildings on the small island. I see a statue on a plateau, a square with a circle on top placed on the farthest reaches of the coastal rocks. Someone has strung tattered nets round the square part of the statue, and these are fluttering in the strengthening wind.

'The widow,' Harvey remarks, as he starts up the engine again. He is pointing at the jetty that has appeared in the grey weather. 'Part of a series of contemporary art works the regional council had installed in various town and country locations up here a few years back. A Frenchman came and put that up here one summer, and then disappeared again.'

'Have you been there often?'

'No, the place has lain empty since the eighties. I called in a couple of times after the Danish guy took over, that's all.'

'What was he like?'

'A competent carpenter,' Harvey replies. 'The first time I met him was one evening when I was on my way home from the farm. I caught sight of him down there by the jetty where he was struggling to haul these bloody windows up to the main building, lugging them on his back, one by one. Triple glazing, frost-proof and God knows what else. Heavy as fuck. Must have cost a fortune. I helped him put them into the bar. Fantastic. No shortage of ambition with that boy, anyway.'

'When did you see him last?'

'A few days before he disappeared.' Harvey runs his fingers over the stubble on his chin. 'Bumped into him in the video shop. He wanted to know what kind of cement I had used for the concrete foundations out on the farm. I think he had reinforcing the jetty in mind.'

I can see the waves beating against the jetty ahead of us. Several columns look rotted through, and the whole structure is swaying in time with the body of water. Some of the ones farthest out have already broken off and are sticking like black stubs of teeth out of the choppy sea.

Harvey peers up at the sky and shakes his head. 'This isn't going to be good,' he says. 'This will be no good at all.'

The clouds are gathering and darkness seems almost to be falling again, even though the day has just begun. A loud creak sounds as the boat hits the rubber strips suspended along the entire length of the jetty.

I feel my stomach churning as I clamber out of the boat and up on to dry land. At once my head is spinning and I clutch the bag I have brought more tightly as I look around for something to hold on to. The next moment it starts to snow: huge thick snowflakes that come sailing down from the dark clouds to melt as they land.

'It looks as if Johannes was right,' Harvey says, as he backs the boat out from the jetty, straight into the steadily mushrooming white flurries.

'About what?'

'That there's a storm brewing—'

Harvey says something else, but the rumble of the engine and the howl of the wind drive his words off in another direction. The very next second there's a loud roar and then the boat is gone.

I clutch my bag still harder as I head towards the main building, with my shoulders hunched against the wind.

Chapter 15

The whole of the small grey island is about to be obliterated by the snow squall that's whipping up with the wind. Sprinting the final stretch to the building, I take out the keys with the yellow tape and inscription *Rasmus Lighthouse & Main Building* from my trouser pocket and insert it into the lock.

The former keeper's residence is a massive wooden house from sometime early in the 1900s, with vertical cladding and windows with white bars. It looks as if Rasmus had made good progress with the renovation of the exterior. The only visible traces of the fire are a few scorched fascia boards, windows and roof tiles, all lying in a heap between the house and the boathouse. The slate roof has been replaced by copper sheets and matching gutters. Even now, in the scudding, oppressive snowstorm, it reflects the light in a strange and somehow beautiful way.

I fumble with the lock until it finally springs open, and hurry inside. The place smells of fresh timber and sawdust, but with a hint of old dust and something else, something indefinable just below the surface. The walls and floor are draped with transparent plastic. Even the upturned Nordland boat, whose hull Rasmus has sawn through to transform into a reception counter, is covered over.

I pull the plastic aside and enter through the door on the right-hand side of the foyer. The room opens out into a large L and appears to have functioned as both bar and meeting room. In one corner sits a curved pink velour sofa with dusty grey curtains piled on top.

In front of that, new boxes are stacked, containing black Murano glass lamps on fine iron stems, white marble tiles and fluorescent pink LED strips, as well as a dozen white bar stools with silver-coloured legs wrapped in plastic.

I take out my mobile phone and key in Anniken Moritzen's number.

'Yes, it's Anniken,' she says sharply. 'Who is this I'm speaking to?'

'It's me, Thorkild Aske.'

I can hear her catch her breath. 'Where are you?'

'I'm at the lighthouse. Standing in Rasmus's bar. Extraordinary place,' I say, even though nothing is finished. The room is full of eighties decor, geometric shapes, rolled-up red rugs, padded bar stools with leopard-print covers and purple wallpaper that has started to bubble at the edges and seams. Even the new picture windows are covered in plastic. 'A bar with a view.'

I cross to the bar counter where Rasmus has set up a camp bed. Magazines about millionaires' yachts are lying on the floor, together with a bag stuffed with clothes and a Rubik's cube. I also spot a box of pamphlets, the pamphlet showing a picture of the place as it had looked when it was newly renovated.

'*Blekholm Conference Centre*,' I read aloud.

'Sorry? What did you say?'

'Apologies. I just found some old brochures from the time when it was a conference centre.'

'Read,' Anniken whispers. 'I want to hear what's in it.'

'OK.' I open the pamphlet and see that Rasmus has circled some of the sentences, as if already planning how the advertising for the place should look. '*We have two conference rooms and one meeting room with a capacity of ten to thirty persons. All laid out so that you can draw strength and fresh inspiration between sessions.*'

'Is that all?' she asks when I stop, hesitating at the part that Rasmus has circled.

'No,' I answer, reading out the final sentence: '*Blekholm offers you experiences to remember long after the conference is forgotten.*'

I catch sight of a trawler in the dim daylight filtering through the plastic on one of the picture windows. It is really a horizontal streak on the blue seascape. Soon I also feel vibrations from the engine rising from the floor where I have planted my feet.

'Tell me more. I want to know what you are seeing, Thorkild,' Anniken continues, tearing me out of the moment. Her voice seems shriller now, as if veering into something else. I have an urge to tell her that I know how it feels when anxiety, fear and panic claw at your innermost defences. To say that it's dangerous to hold it in for long, and that one day it will be forced to burst out. But I don't dare. I'm not the sort of person who can be there for you when you need me.

I put away the pamphlet and squat down beside the camp bed. 'A bed with some clothes, books, a washbag and a few razors in a white plastic container.' I draw aside the sleeping bag on the camp bed and spot a handheld Motorola radio transmitter and two packs of spare batteries at the foot. 'Some kind of maritime VHF radio, I'm not sure.'

'What else?' she presses me harder as we continue the verbal balancing act in which every word, every single breath in and out, tugs at the fibres of the gossamer fabric that keeps her away from the abyss. 'What do you see?'

'Nothing more,' I finally tell her, leaning my back against the bar. I close my eyes and try to pinpoint the sound of the diesel pistons on the trawler out there in the storm. 'He's not here, Anniken,' I whisper. I notice how dizzy and tired I am, and don't have the strength to continue with this painful exercise. 'Rasmus is no longer here.'

'But he *was* there,' she answers harshly. 'Very recently. His scent is still there – it's just that you don't recognise it because you've never held him close to you as I have. I can picture him where you are now, that's what is so difficult. And I just can't, I can't bear to—'

'I'll keep searching,' I whisper as I stand up, mustering the energy to continue. 'Stay on the line, Anniken, and we'll move on. OK?'

'The lighthouse,' she exclaims with renewed vigour in her voice. 'He used to phone me from the top of the lighthouse.'

'Fine. Let's go there and look.'

An oppressive, subterranean darkness follows on my heels in the whirling snow as I dash across the courtyard towards the steps that lead up to the lighthouse. Even though it is still early morning, it feels as though I'm in a waiting room midway between night and day.

The steps are carved into the mountainside, with a rusty iron frame used as railings. The lighthouse is an eight-sided cast-iron cupola with a broad, red-painted stripe just beneath the dome. The red and white concrete tower practically merges with its surroundings.

'This is going to be a suite,' Anniken begins to say once I have closed the outer door and set off up the iron steps that spiral to the top. The walls are of polished concrete with black and white pictures of stormy seas and dark clouds all the way up: an outstanding way to create an ambience for tourists. 'A location between sky and sea with a vista in all directions.'

'I can see it.' The innards of the lighthouse tower are stripped out, the whole lens apparatus removed and an antique four-poster bed placed in the centre of the room so that you have a 360-degree outlook through rows of new windows. LED strips are mounted on the ceiling and the beginnings of a ship's deck looks as if it will form the flooring.

'Isn't it lovely? Rasmus sent photos of the room and the view from his mobile phone.'

'Beautiful,' I answer, as I push a few boxes of wall tiles over to one of the windows and sit down.

'And the view? What do you think of the view? Rasmus calls it unbeatable.'

'He's right.' I lean my elbows on the window ledge and let my eyes wander out to the wall of snow drifting past in a steady stream. It is so grey that I can no longer see land. Even the instrument shed at the lighthouse entrance has vanished in the dense grey mist. 'Absolutely unbeatable.'

'Thanks.' Anniken exhales noisily.

'Anniken,' I venture after a lengthy silence in which we both just listen to each other's breathing. 'I don't know what else I can do.'

'Just come home,' she replies wearily. 'I know it now, I do know. He's not there. Oh God,' she gasps, as her defences finally collapse. 'He's not here any more ...'

I stay seated, my mobile phone pressed against my ear, long after Anniken has hung up. The trawler has gone, the wind is howling outside and the waves beat on and on against the shore. In front of me, soft flakes of snow spiral effortlessly, elegantly, like couples in some exotic dance.

I think of Frei.

CHAPTER 16

Dancing class with Frei, Stavanger, 24 October 2011

I saw Frei again the very next day after our meeting at Café Sting. She swept through a crowd of youngsters and their parents, all with painted faces and weird hairdos, into the Sølvberg Arts Centre. According to a sign at the entrance, a family festival was being held all day long.

A stage had been rigged up inside the foyer. Costumed mythical creatures with black eyes and smiling, predatory faces were leaping and dancing around the audience holding blue UN flags in their hands.

It was far from coincidental that we met there. I had been in the area for several hours already, drifting restlessly between the arts centre, exhibitions and shop windows as I waited for the time to near six o'clock. As stated on a notice on one of the second-floor glass doors, right beside a course in origami for children, ballroom dancing classes were held on two days a week.

In an effort to make the time pass more quickly, I had even watched a theatrical performance about a boy with a soft toy snake that came alive only when they were together, but it did not help much.

Frei was dressed in dark-grey leggings with a hoodie and black trainers as she jogged up the stairs without spotting me. At my back, a microphone switched on and an enthusiastic woman announced that it was now time for the troll and sprite workshop in the arts centre basement.

I smiled apologetically at a woman dressed as a witch when I narrowly avoided bumping into her daughter, dressed for the occasion in a hand-knitted bumblebee costume with two matching badminton rackets taped to her back. I then continued to the stairs.

On my way up, I passed a number of display stands at the library entrance on the first floor, where courses in ceramics and henna art were offered. Rapid drumbeats and rhythms began to drift from the foyer below.

At the door to the dance venue I took time yet again to curse my own almost perpetual childishness and lack of ability or will to do anything about it. Then I stepped inside.

I entered a crowded corridor and could see Frei's black trainers and hoodie in the chaos of clothes, scarves and footwear. A glass door divided the corridor from the dance venue where the class, of six couples and an instructor, was in full swing.

Frei was dancing with a slim, well-groomed man with thick, slicked-back, black hair. His hair gel was almost luminous as they glided around in each other's arms, floating across the floor, guided by intertwined body movements in perfect step and choreography.

'Elegance and lightness, folks,' the dance teacher called out in broken Norwegian, clapping her hands and manoeuvring herself flexibly between the couples. 'And twirl.' A red-haired man in his forties stood in front of a stereo system with his hands folded under his chin, gazing dreamily at the dancers.

'More twirls. More, more, more!' the instructor said, clapping, before she turned abruptly to her red-haired assistant in the corner and beckoned him over. 'Señor Alvin, come!'

Barefoot, Alvin tripped across the floor and into the woman's open arms. She placed her right hand round his waist, extended her left hand to one side with her elbow bent and palm upturned, and led him round in a semicircle through the other dancers. 'Come, now, *temperamente*: forward, to the side, together. Back, to the side, together. Come on now, one, two, three, four, five, six! All together: one, two, three, four, five, six!'

Behind me the door swung open and a plump woman came in, panting. 'Damn, have they started?' she gasped, squeezing between the door and me, and peering in.

She kicked off her trainers and struggled out of her multicoloured poncho-style outer garment that resembled a patchwork quilt. 'So,' she said, smiling, puckering her lips and winking at me: '*Bailar pegados!*' She wrenched the door open and slipped sideways into the room. 'Alvin! Alvin, sweetheart, sorry I'm so late.'

Fleetingly, Frei and her dance partner with the shining hair slowed their orbit, exactly in time to see the woman and Alvin reunited in a passionate embrace on the dance floor, just as the glass door, which until now had closed neatly and tidily with the help of an electric mechanism, sprang back and slammed straight into my fingertips.

Pain shot up my arm and it was all I could do to push the door back far enough to release my fingers so that I could wheel around and make for the exit and stairs. My bloodshot fingers pumped and pulsed as if they were sticks of dynamite about to explode.

'Thorkild!'

'Oh no,' I groaned, stopping at the first step. Hiding my throbbing hand under my jacket, I turned around.

'Frei.'

A burst of cheering broke out from the foyer to introduce the intrusive rumble of drum rhythms accompanied by African musical instruments.

'What are you doing here?'

'Important police work,' I answered, pressing my arm against my aching fingers beneath my jacket. 'An undercover job.'

'Really?'

'Absolutely. Haven't you heard?'

'Heard what?' she asked as the percussion instruments lifted the music to fresh heights while the audience stamped their feet and clapped hands to the beat.

'The Origami Killer is on the loose in the city again. Grisly case. Truly horrific.'

Frei shot a look in the direction of the wide-open door behind her where a small Asian boy of seven or eight sat folding paper figures along with three girls of about ten and their parents, before she turned to face me again. She was standing beside a pillar just outside the door into the dance class, meeting my eye with the trace of a smile.

'Is your hand OK?' she finally asked.

'Super.'

'Shall I drive you to A&E?'

'No.'

'OK. Are you coming with me?'

'Where?'

'In to dance.'

'No, I—'

'Don't you want to?'

'Hell, yes, but, I mean, what will your partner say?'

'Robert?' Frei giggled. 'But I told you about Robert at Café Sting. Uncle Arne's boyfriend? Fabulous dancer, probably an insatiable lover too, don't you think?'

'Bound to be,' I muttered, feeling the pain in my fingertips sneak out through my pores and into the dance rhythms that filled the arts centre from basement to roof.

'Come on,' she said, holding her hands out to me. 'I know you want to.'

I no longer heeded the pain in my fingers as I walked with Frei back to the dance class. All I was aware of was an intense tingling from the point where the skin of our hands met, and the scent of her hair. If I had leaned my head closer, her curls would have brushed my cheek and lips.

CHAPTER 17

I am startled by a knocking on the door at the base of the lighthouse. It makes me lift my head from the window ledge and stare straight into the condensation from my breath that has settled on the glass. I'm freezing: my clothes are cold and my boots feel stiff and tight, as if they have shrunk while I slept.

More sounds of knocking come from somewhere below, hollow thumps that vibrate up through the building all the way to where I am sitting, and make the plaster rattle at the joins. I wipe the mist from the window with my sleeve and peer out: the weather is just as grey as before, and you can't tell day from night. A constant stream of snowflakes drifts past, and waves roll towards the tiny island.

All at once I catch sight of a man hovering between the instrument shed and the front door of the lighthouse. He is wearing oilskins and waving at me with one hand while pointing to the sea with the other, as if trying to point out something in the blizzard.

I am about to lean closer to the glass to see out more clearly when I am startled to hear loud bangs from the base of the lighthouse. The next moment everything is totally silent again, as if I have suddenly found myself inside a vacuum. I turn back to the window and look out: the man in the oilskins is gone.

Between the boathouse and the jetty, a gap has opened up in the storm in which the snow is spinning like a miniature tornado high above the coastal rocks. I can only just make out something protruding from clumps of thick seaweed eddying at the water's edge below: a shape whose colours and structure do not belong in the seascape.

Standing up, I am on the verge of walking to the door when there is another bang from below. The echo of metal against metal sings through the walls as the wind plays havoc, whistling and gusting outside. I approach the door and tentatively turn the handle.

A chill blast of raw, salty polar wind rips through me into the room when I open the door. I cross to the stairs, where I lean over the railings and try to see down through the spiral staircase to the source of the knocking. 'Hello!' I shout so loudly that my voice cracks, and brings on a violent burst of coughing.

No one answers, and I start to make my way down the stairs. Now and again the door below slams so hard against its frame that the whole staircase shakes and I have to stop and brace myself against the banister as the din assails my ears.

The entrance is covered in snow and the metal door stands open. I grab the door handle and begin to use my feet to shovel away the snow before stepping outside and closing the door behind me.

The shape to the south-west has disappeared, and there are no footprints in the snow. I run down the steps, holding tightly to the railings. All around me the snow is cascading into the sea and battering the island, while waves spray the rocks and rotten jetty with frothing foam. A layer of new snow covers the courtyard. The wind will soon escalate to reach gale-force.

Halfway across the courtyard, I come to a halt. There is no sign of anyone else having been here recently, but something catches my eye out in the water. The wind builds up as I approach the sea. The snow is falling less densely now and I can see the lights of houses on the distant shore.

I drag myself forward for the final stretch, finding handholds on tussocks of grass and rock crevices to avoid sliding or falling. The lights on land shine yellow like lanterns on an old ship, and then the snowstorm intrudes once again between island and mainland, making everything shades of white and grey.

Drawing breath, I feel the taste of salt fill my mouth once I reach the water's edge. The sea here is deep and black with white crests that strike the rocks, causing the body in the water to rock from side to side.

The dead body is floating face down. Only the back and a few slimy strands of hair are visible; the rest is underneath the mass of water or hidden in the tendrils of seaweed. I am on the point of leaning forward to seek for a groove or shelf to take me closer to the body when a huge wave surges in front of me. I just manage to fling myself backwards before the wave takes hold of the corpse and washes it away over the rock where I had been standing a moment before.

'What on …' The body is left lying on its back directly in front of me and then begins to slide down the slick bedrock towards the water. In the end it rolls over the edge and flops down. Soon the torso breaks the surface again between the clumps of seaweed to settle face down with arms dangling at the sides.

Gasping, I take a deep breath before launching myself on my stomach over the rock to the water's edge, eyes peeled for further waves. I inch my way forward, supported by shells and barnacles on the stone, until I am so near that my fingertips can touch the body. Soon I am able to haul it towards me and up on to dry land. I get to my feet and start to tow it behind me across the rocks to the boathouse.

'I don't understand,' I pant, out of breath, when I finally come to a standstill and drop the body. 'It's not Rasmus.' Wet and exhausted, I sink to the ground beside the cold corpse. 'It's not even a man.'

CHAPTER 18

'Bjørkang,' the voice answers at the other end with a mixture of curiosity and exaggerated authority. Bjørkang sounds intoxicated. In the background, someone is playing an accordion.

'Thorkild Aske here. Did I interrupt something?'

'It's Saturday, a meeting at the accordion club.' He stops for a second before adding: 'Who did you say you were?'

'The private investigator.' I can hear the local police chief talking to someone in an undertone, and the accordion music stops. I block my other ear with a finger to shut out the wind. 'I'm out at the lighthouse—'

'But what the hell are you doing, still out there? Haven't you seen the weather forecast? There's going to be a gale tonight.'

'Yes, I know, but—'

'Are you on your own?'

'No,' I reply, standing up. I begin to skirt around the corpse on the ground in front of me, all the while struggling to get my circulation pumping and keep the cold from capturing my body.

'That's fine, then,' he says, exhaling loudly. 'Make sure you both get away from there as fast as fuck before the storm gets hold of you.'

'I found a body,' I start to say. 'In the sea.'

'The Danish bloke?' I can hear whoever is with him stop mumbling, and they both fall totally silent.

'Is there anyone else reported missing up here?'

'No—' Bjørkang begins to say.

'It's not Rasmus,' I interrupt. 'It's a woman.'

'Can you say that again? You found what in the sea?'

'A woman. She looks as if she's been stuck on the seabed for some time.'

There is a lengthy pause, broken only when the local police chief's deep voice resumes. 'Are you sure?' While I stand there phone in hand listening to the policeman's heavy, rasping breath, my eyes survey the dead woman beside me.

Her body is bloated, distended like a beach ball in human clothing. The corpse has no face, just mid-length hair and scalp left after the time spent underwater. One arm is torn off at the elbow joint and the entire lower jaw is replaced by a dark hole where the white cartilage rings of the windpipe and pink oesophagus protrude. The grey tongue muscle hangs between her breasts like a necktie under the grotesque hood of flesh encircling the skull.

She looks young, and is wearing a flimsy nightdress and a T-shirt on top, with a picture of a horse emblazoned on the front. Blood has collected in her legs, all the way up to the knees as well as in her neck, which is deep purple. Her arm and arm stump, in addition to the upper part of her chest, are marked by bronze post-mortem lividity. The bobbled, leathery skin resembles goose flesh.

'Hello, Aske! Are you there?' The voice at the other end sounds hollow and crackles as a deeper bass tone mingles with it when the two men I am speaking to raise their voices, sounding jittery.

'Yes, I'm here.'

'Are you sure that—'

'Can't you just come here, Bjørkang?' I insist, feeling irritated. 'This place gets under your skin and I want to be out of here. Come as fast as you can, for fuck's sake!'

'OK. I'll requisition a boat right away and then I'll phone you back very soon. Agreed?'

'Agreed.' A crackling noise, and then he hangs up.

I continue to stand with the mobile phone in my hand, without taking my eyes off the faceless woman. The snow has formed a white cobweb film across her, making her look like a morbid version of a snow angel, with only one and a half wings.

I walk over to the pile of materials between the main building and the boathouse and start to carry roof tiles and scrap metal over to place in a circle around the dead body. Afterwards, stepping inside the foyer in the main building, I tear off the plastic sheeting in front of the restaurant door and bring it outside.

I wrap the corpse in the plastic, before placing the tiles and metal on top. Soon I have finished and head off up to the lighthouse again.

I jog breathlessly up the steps and close the door behind me before continuing up the internal staircase all the way to the top, where I resume my seat by the window. Not long afterwards, my phone rings again. It's Bjørkang's colleague, Sergeant Arnt Eriksen, sounding out of breath.

'Aske, is that you?'

'Yes.'

'Bjørkang just called,' he begins. I can hear that he is outdoors. I hear gravel crunching as he walks, and in the background a woman's voice speaking on another phone.

'Where are you?'

'On my way down from Sørøytoppen. My girlfriend and I were making our way up to the cabin there when Bjørkang phoned.'

'Are you coming to fetch me?'

'Yes,' Arnt says. 'Just take it easy. I'll pick up Bjørkang from his house as soon as we get down. He's sent the accordion club boys home and put the cognac back on the shelf. Ha!' His laughter sounded artificial and forced.

'Why are you phoning me?'

'I just wondered …' he begins to answer.

'Wondered what?'

'Well, Bjørkang said you had found something in the sea. A … woman?'

'That's right.'

'What does she look like?'

'What do you mean? Do you know who she is?'

'What? No,' the sergeant groans. The female voice at his side is speaking more quietly now, as if she too is inquisitive about why Arnt is phoning me. 'I just had to check whether he had heard right. I mean, we're not missing anyone apart from the Danish guy here, so—'

'So what, then?'

'No, forget it. Just take it easy, we're coming.'

Arnt Eriksen hangs up and my gaze returns to the little island. Far below on the shore I see the plastic bundle with the dark bulk inside, as snow flurries pass just above. My cheek is throbbing and I tear two OxyNorm capsules from the blister pack. I certainly have no wish to stay out here in the storm on my own. The thought of the woman beneath the tarpaulin, Rasmus, the bar, this lighthouse and this whole island gives me a sudden overwhelming urge to retch. 'I should never have come here,' I whisper to myself, tossing the capsules down my throat. 'This isn't going to end well, that's for sure.'

CHAPTER 19

'I thought it was her – Frei,' I say when Ulf finally picks up the phone. I've been waiting up here at the top of the lighthouse for ages, without seeing or hearing any sign of Bjørkang, Arnt or the boat they are supposed to bring to fetch me.

'Where are you?' Ulf asks, lighting a cigarette before exhaling noisily.

'I'm on that island, the one belonging to the Danish guy. I've just hauled a woman out of the sea. When I saw her in amongst the clumps of seaweed, I thought for a minute it was Frei.'

'Why?'

'Why? What the fuck do you mean, didn't you hear what I said? I just found a woman with no face in the water!'

'Yes, I hear you,' Ulf answers. 'But you must have come across dead bodies before, haven't you?' I can hear the tension in his voice, even though he's trying to hide it.

'Yes, of course,' I reply, taking a deep breath. 'I just wasn't prepared for it.'

'Who is she?'

'Don't know. It's a young woman, maybe early twenties. No idea who she might be. The local police chief says there's no one reported missing.'

'Is she with you right now?'

'No. I wrapped her up in plastic sheeting and left her lying down there on the rocks.'

'Fine.' Ulf sucks hard on his cigarette, purring in rapture between puffs. 'What if we talk about something entirely different while you're waiting to be picked up? Because you'll be picked up soon, won't you?'

'Yes, I've alerted the local police. They're on their way.'

'Great. Why not start by telling me why you thought it was Frei? After all, she's buried in a churchyard out in Tananger.'

'I just … I'm stressed out,' I finally tell him, taking a deep breath. 'I just took two of the Oxys, and they haven't started working yet.'

'Soon,' Ulf intones, without a trace of anxiety, obviously enjoying his cigarette. 'Very soon, you'll see.'

'No,' I complain, as my eyes are drawn to the grey-black cloud cover above me. 'Nothing works up here. Even the sky is shut behind a wall of darkness and snow.'

'The sky?' Ulf pulls up short. I can hear him holding the smoke inside before exhaling steadily again. 'What is it about the sky, Thorkild?'

'Nothing,' I answer crossly. I know I've said too much. 'I spoke to Liz, as you wanted me to,' I add in an effort to change the subject.

'That's good,' Ulf responds quietly and mechanically. He lets me gabble on, waiting for a gap in the conversation, when he can twist the dialogue in a direction I don't want it to take. 'Did that go well?'

'I punched Arvid.'

'Why?'

'He's a scumbag.'

'Undoubtedly. Did it help?'

'Absolutely.'

'OK, Thorkild. Did you get time to have a chat about your mother and father?'

'No.'

'Well, maybe another time.'

'Christ.' I grit my teeth so ferociously that the pain in my cheek shoots all the way up to my eyeballs. I squeeze my eyelids tight shut and try to conjure up the medications that can help me away from this ice-cold, pain-filled space, into somewhere else. 'She doesn't want to come,' I sob, clutching my mobile. 'Not in this weather. Nothing works, I've already told you. I can't even take a shit up here.'

'Listen to me, Thorkild. What if we just take a breather now, you and me, calm down and have a conversation while you're waiting for the effects to kick in? Then you can buy something for your stomach when you arrive back on dry land. Duphalac. OK? Can you do that? Will you?'

'Yes,' I sigh. 'I can do that.'

'Brilliant. I'll light a cigarette and we can start all over again? OK? Is that fine?'

'Fine.'

Ulf lights a cigarette and inhales: I can hear the crackling down the phone line. He is holding his cigarette, filling his lungs right down to the deepest reaches of his bronchial tubes, until he breathes out with sheer enjoyment, heaves a sigh of contentment, and whispers: 'OK, Thorkild, let's talk.'

The snowflakes drifting past outside dance up and down on the air currents: sometimes they come to a virtual stand-still, as if poised in free fall until another gust of wind catches and chases them on through the night. Heavy waves pound at the island from every direction. 'Have I told you about when I met her again?' I ask, resting my cheek on the cold glass where the vibration is stronger.

'Many times, Thorkild. Many times.'

'All of a sudden she was standing there in the jets of water.' I feel an impulse to laugh, even though my body is trembling with cold. At the same time, I notice something give way inside me. Threads slipping away from one another. Nerve pathways and muscles stretching out to their full extent, led by tiny white snowflakes dancing around in my belly, bringing oxycodone with them, lifting it up through my blood vessels to my head and the pain receptors waiting in there. Every particle settles in the right slot and puts me back together inside, fragment by fragment. 'It was her, but all the same not her, do you understand? Her, the way she is now. That was why I knew it wasn't a dream.'

'Tell me instead about why you went down to the showers, Thorkild. Tell me about your meeting with Robert, Arne Villmyr's boyfriend. I know he visited you in prison, that you had a conversation that day. What was it he said to you?'

'I knew her at once,' I plough on, ignoring his questions. It is snowing, not only outside, but also inside. Not scudding and driving like a raging storm, but strands of soft crystals, big as goose down, floating down over me in a perfect sensory experience such as will only occur in a properly joined-up brain.

'He told you about Frei, and someone else?'

'No,' I break in. Some sort of imbalance is beginning to spread through my body, and I feel anger building up inside. 'I don't want to talk about that now,' I say in annoyance.

'We haven't talked about it yet,' Ulf purrs with equanimity, still puffing on his cigarette. 'Maybe this would be an opportune moment to examine it together while we're waiting for that boat?'

'No!' I snap into the receiver.

'OK, Thorkild, I didn't mean to push you.'

Outside, it will soon be completely dark. Thin streaks of pale light dart down through the snowstorm, striking the island and the sea at chosen points, and the rest is blue-black, white or shapeless and grey. 'Shh.' I place my hand over the microphone. My gaze is fixed on something all the way down there beside the jetty.

'What is it?' I can just hear Ulf's voice through my fingers. The light sways to the beat of the ocean's surface around the jetty, where a human figure is pulling itself up out of the waves. As soon as he is out, he stops to look around, pausing, his hands on his knees, as if catching his breath.

'I have to go,' I whisper just as the figure stands up to its full height and sets off towards the boathouse, where the body of the faceless woman lies rolled up in plastic. 'Someone's here.'

Chapter 20

I gather up my belongings and hurry downstairs to the base of the lighthouse. Outside, a gale is blowing. The wind has grabbed a torn-down radio mast and is tugging and tearing at it, to and fro across the concrete foundations. The grind of metal on concrete is ear-splitting.

I see no one, and no boat down by the jetty where black shadows twist and turn in the wind, while the snowstorm rages, imprisoning us on this grey, fog-bound rock. A fresh snow squall rolls in over the island as I take hold of the railings and start to haul myself down the concrete steps towards the conference centre and the boathouse.

The corpse is no longer where I left it. Only an impression in almost transparent slush shows where the head and torso had recently lain. While I stand there looking down at the imprint of an angel with only one wing, there is suddenly a splintering crack from the jetty.

I turn and see surging breakers sweeping in across the island. The rotten jetty structure swings noisily up and down with the waves. All at once I catch sight of the figure again in the middle of the jetty where metre-high waves are soaring high in the air and crashing down again. It is a man whose body blends almost completely into the dark background and the blizzard rampaging between us. He is wearing a diver's suit and mask and seems to be dragging the body of the faceless woman behind him to the edge of the jetty.

'Hey!' I wave my arms as I take a few steps forward. 'Wait for me!'

The man stops momentarily and looks up at me through the driving snow. He stands gazing at me and then adjusts his grip and continues to drag her onwards.

'Wait, for God's sake!' I start to jog down to the jetty. I can see they have reached the edge. He crouches forward and looks down at the water as if considering jumping in.

Once again I hear the pier columns crack. The whole jetty is rocking violently as it creaks and bangs against the metal underpinnings. The man in the diver's suit is now standing at the far edge of the jetty with the body of the faceless woman in his arms. Motionless, they just stand completely still as the seawater lashes over them.

Yet again the foaming surf comes crashing over the teetering jetty. The water surges into the air and over the structure, followed by fresh grinding from the foundation pillars. The next time I look at the place where they'd been standing, they are gone.

'No!' I cry, and totter shakily forward before stopping again. The jetty is about to tear loose from the bedrock. The very next minute, still more strident cracking noises issue from the pier foundations before the whole jetty lifts right out of the water. All the time, there are groaning noises as timber fractures, brackets snap and reinforcing irons break off. The construction rises vertically out of the water before flopping heavily on to its back and starting to slide away from the rocks, out into the dark depths of the ocean.

Suddenly I glimpse them again, not far from the broken jetty. The water is shallower there, paler than farther out, and I can see two shadows, one floating head down so that only a tiny part of the back breaks the water, while the other floats just below the surface, as if watching me through the murky layer.

I linger there for a few seconds before breaking into a run across the rocks to the shore. At the same time, the man lugs the dead body away from the island.

'Hey, where are you going?' I yell in desperation. 'Where's the boat?'

But now I skid and lose my footing on the slimy ground. I crash down and smack my cheek and head on hard stone, ending up half wedged between two rocks no more than half a metre from the water's edge.

My leg is turned in towards land and my head is safely cushioned by the seaweed, with seawater gurgling its way up between the rocks from below and washing over my face. Shivers course up and down my spine as I struggle to break free.

'What the hell just happened?' I gasp, exhausted, when at last I have hauled myself up on to safer terrain. I am on the rocks, crouched on all fours like a soaking-wet dog, gasping for air. The jetty has vanished into the darkness. Nothing remains behind me but loose fragments of concrete, clinging to the mangled, rusted reinforcement bars sticking up out of the bedrock. The underwater shapes are gone too, as if they were never here and I have just woken from one of those dreams that feel almost real.

I lie flat on my back with my face turned to the sky. Snow flurries hurtle past and the darkness has settled like a lid over the island. Eventually I notice a faint rattling from the boathouse door not far off.

Mustering all my strength, I strain to magic away the pain in my face and head until I can get to my feet and make my way in that direction. I hunker down in front of the door and make an effort to peer in underneath it through a chink at the bottom.

I stand up again and use my fingers to fumble for the padlock. A new padlock is fastened to the door, and I walk over to a pile of materials and pick out a piece of metal flat enough to wedge between lock and door.

The timber creaks as I press the metal down and pull it towards me until one of the planks breaks loose, along with four thick nails, and the whole door bursts open, almost propelling me down on the pebbly shore as the wind takes hold of it. I squeeze into the boathouse without even trying to close the door behind me.

Once inside, I can make out the contours of a brand-new standby power supply system in boxes, with only the exhaust unpacked and placed in the middle of the floor, beside several metal lighting masts connected to an outdoor jacuzzi. Five brand-new diving suits hang from hooks along one wall.

The length of transparent plastic sheeting that I had used to wrap the corpse has blown in under the door and attached itself to the standby power supply exhaust system. The wind tugs at the plastic, making it rustle and scrape on the motor parts and the concrete floor. I walk further inside and bend down to pick it up.

I lift the plastic closer to my face, aware of the foul smell of decomposing flesh, skin, muscles and intestines that lingers in the plastic folds. The impression of the dead body is still visible, where flaming yellow threads of blood and body fluids form the outline of a human being.

I fold up the plastic and jam it tightly under one of the exhaust system pipes. When I phone Bjørkang no one answers.

Neither Arnt nor Harvey picks up the phone either, and soon after that, my mobile runs out of battery. I decide to wait here in the boathouse until help arrives, and find some packing cases against the rear wall to clamber on to. I'm chilled to the bone, and have a splitting headache from my fall on the rocks. My body aches, my cheekbone throbs, and my legs feel numb.

I fish out the water bottle from my rucksack and lodge it between my legs as I scour my pockets for the bag with my blister packs, loose tablets, dispenser and capsules, then toss the pills down my throat and lean back against the wall with my arms hugging my body. I close my eyes and try to focus on the wind outside.

SUNDAY

CHAPTER 21

Somewhere beyond the boathouse I can hear the roar of a boat's engine. I slide my fingers out of my jacket sleeves, scramble down from the packing case and walk outside. Here an ochre-yellow fog has materialised through the sporadic snowflakes still swirling in the air. The thudding of the outboard motor grows louder, and through the snow I can see an old man in a motorboat heading towards the lighthouse.

'So you're still alive?' Johannes yells. His boat slides into position, parallel to shore.

'Only just,' I say, trembling with cold, and stand still in front of the concrete remnants of the jetty demolished by the storm overnight, while Johannes works the boat closer.

'What the hell?' he exclaims when he catches sight of the stumps left behind by the jetty. 'Where's the rest of it?'

'Disappeared some time last night.'

'Harvey contacted me on the walkie-talkie and asked me to come out and pick you up as soon as the weather permitted. A number of the mussel poles had come adrift during the storm and were floating out to sea. He's still out there. They say there's going to be another gale not long after daybreak, so we're short of time.'

'Have you seen anything of the local police chief?'

'Bjørkang? No, why should I? It's only half-past four, you know. The servants of social democracy don't get up for another few hours yet.'

'I spoke to him last night,' I tell him, climbing aboard as soon as he comes alongside between the projecting reinforcement bars and heavy chains rattling roughly at the water's edge. 'They were to come here as soon as they got their hands on a boat.'

'Well, maybe they had more important things to do.' The old fisherman starts to reverse the boat out. The skies above us are darkening again. 'You'd probably have managed there for a while yet, if you had to.'

'I found a woman in the sea.'

'Oh?' Johannes spits through his front teeth as he adjusts the steering and puts the boat in gear. The boat starts to work its way forward across the choppy seas at a halting speed. 'Are you sure it was a woman?'

'Yes.'

'Where is she now?' he asks, seemingly unruffled, as if I'm talking about an unloved hamster or a stranded goldfish.

'Someone came up out of the water and took her away.'

'Who?'

'A man.'

Johannes nods dourly to himself as he steers the boat around a huge drift of seaweed and plastic debris swirling agitatedly around on the swell below a rapidly changing sky.

'The Sea Ghost is a corpse without a face,' he ventures without preamble. 'A dead fisherman who drifts around on the sea or sails in a half-boat with tattered sails. A warning of death and despair.'

'Does it ever happen to dress up in a diving suit and crawl on land to return the dead to the sea again?' I try to laugh, but the laughter sticks in my throat. The wind grabs the last ragtag snowflakes and starts to chase them across the surface of the sea.

'No,' Johannes answers. 'He doesn't do that.' His face is hard and there is a network of purple veins just below his skin.

'Didn't think so.'

It is five o'clock when the boat smacks against the pebbles on the beach at Skjellvik. I jump on to dry land and help Johannes to drag the vessel up and into the boathouse. We begin to walk along the road to the far end of the bay, where an old house with flaking paint sits in the open, exposed to wind and weather from the mouth of the fjord beyond.

'Do you have a mobile phone I could borrow?' I ask, once Johannes has kicked off his boots in a hallway that smells of cod liver oil and honey, and pulled on a pair of thick grey socks. 'Mine's run out of power.'

Johannes points. 'Use the landline in the living room. That's cheaper. And I'll lend you a battery charger.'

The parquet flooring sticks to my sweaty socks. I put my own phone on to charge before locating the landline and calling Bjørkang's mobile number.

It is switched off.

'Do you have the number for his sergeant, Arnt something-or-other?'

'Eriksen. Arnt Eriksen. Look in the phone directory.' Johannes hands me a local directory and then sits down in a chair, covering his legs with a tattered patchwork quilt. 'But if you wait for a while, we can contact them on the walkie-talkie instead.'

'The walkie-talkie, yes, good idea,' I mumble. 'Rasmus had one too.'

'Who?' Johannes looks quizzically at me.

'Rasmus,' I repeat. 'The Danish guy out at the lighthouse.'

'Yes, that's the simplest way. Costs nothing, you know,' he nods. 'Completely free.'

'Radio hamsters – screech, scratch, crackle!' I say, taking a seat on a burgundy sofa from the fifties.

'What did you say you were?'

'Forget it.'

Harvey is first to answer on the shortwave network. He is still out at the mussel farm and says he'll come as soon as the mussel poles have been secured.

After I've spoken to Harvey, I phone the police in Tromsø. They haven't heard anything from Bjørkang, or about any boat requisition or picking up someone from a lighthouse during the night. I explain about the woman in the sea and the person at the other end asks me where she is now. I tell him that someone in a diving suit came up out of the sea and took her with him back underneath the waves, and the policeman sighs and asks me to call the local police office when it opens. Then he hangs up.

Johannes goes out into the kitchen and returns with a mug of coffee for me, and takes his with him when he goes out into the hallway to get dressed. 'It's blowing up again,' he says from the hallway. 'I need to go down to the shore and secure some of the boathouse doors, but I'll be back shortly.'

Pulling my feet up from the cold floor, I lean back in the hard, old-fashioned sofa and try to find some kind of restful position. 'Do you think they set out last night, and something happened to them in the storm?'

'No. Just lie down and relax for a bit, won't you?' Johannes says placidly. 'You look as if you need some rest.'

I can see in his eyes that he is not as calm as he is keen to appear, while he stands in the doorway gazing from me to the sea outside the window. In the end he turns on his heel and disappears through the front door. I fall asleep almost at once.

Chapter 22

On the radio, a local weather forecaster announces that a gale is expected, with gusts of up to 140 kilometres per hour in exposed areas throughout the afternoon. People are asked to stay inside during the storm and avoid traffic and bridges if possible. The long-range forecast predicts a fresh area of precipitation moving northwards. The intense low pressure, combined with the full moon, makes a spring tide likely, with an estimated water level of up to one and a half metres above the norm.

'Yes, that's something,' Johannes says when I open my eyes and look around. He has placed the walkie-talkie on the table beside that day's edition of the local newspaper, together with a greasy paper bag of doughnuts. 'The last time there was a spring tide I had to pump out the cellar and dig up all the drains around the house. A real brute. And impossible to set out nets. They fill up with all sorts of shit, you know.'

'What time is it?' I haul myself up and put the cup of cold coffee on the table.

'Eleven a.m.,' Johannes says. 'You were miles away. Didn't want to wake you – it looked like you needed the rest.' He reaches to take a doughnut before pushing the bag across to me.

I help myself to one, dip it into my coffee and take a bite. My mouth feels numb. The taste of coffee rouses the tiniest taste buds on my tongue to tell me how foul this is, at the same time as my stomach reminds me that something needs to get out before anything more can get in. 'Has anyone phoned?' I ask, putting the doughnut back on the table beside the cold coffee.

'No,' Johannes answers. 'But I spoke to Harvey a while ago. He's on his way over.'

'And the police chief?'

Johannes bites a doughnut in two. 'No.' He crams the other half into his mouth and puts the steaming coffee to his lips. 'Nothing.'

I switch on my mobile and see that Ulf has called five times. There is also a text message from Anniken Moritzen, asking me to visit her at her office when I get home to Stavanger. I pour out a fresh cup of coffee before I phone the local police station. A voice message there recites opening times and explains that all enquiries outside those times must go to police headquarters in Tromsø.

But at police headquarters in Tromsø there is no one who can tell me where the local police chief and his sergeant are to be found, and the woman at the other end has no wish to join me in wondering why neither of them is in the station or answering the phone.

'Fucking idiots.' I put the mobile on the table.

'Mhmm,' Johannes smacks his lips, wiggling his toes inside his woollen socks in sheer delight and hugging his coffee cup as he uses his tongue to clean his mouth. 'There are so many of them too,' and he smiles with amusement, blinking so that his bushy eyebrows bristle like spines on a thorn branch. Johannes's mop of grey hair is cut short at the nape and combed back over the crown. He looks like an adventurer of the type you see in black and white photographs, the ones who ventured out to discover all sorts of things in Arctic regions before the world wars.

'Many of them?' I ask tartly, sipping my coffee. 'Police or idiots?'

Johannes is on the point of answering when there is a loud knock at the front door, followed by heavy footfalls in the hallway. The next minute, Harvey appears at the door with a forced, weary smile on his face.

'There you are, then.' He rubs his hands together, as if to get the circulation going.

'There's coffee in the kitchen.' Johannes draws the bag of doughnuts towards him. 'And doughnuts.'

Harvey disappears out to the kitchen. 'You made it ashore,' he says on his return. He picks out a doughnut and sits on the sofa beside me. His clothes are damp, and tufts of hair stick out in all directions. His face is greyer than I can remember, and his lips narrow and bloodless.

I give a silent nod.

'What a night,' Harvey comments, shivering. 'Some of the mussel poles had drifted almost all the way across to Steinholmene on the other side when I got hold of them. Had to just tie them to ones that were still standing, and weigh them down with an old millstone I found on land. I'd rather sort out all the mess when the weather improves.'

'The southerner says he found a woman in the sea last night.' Johannes luxuriates, stretching his toes out again, before getting to his feet and crossing the room to switch off the radio.

'Oh,' Harvey grunts. 'Are you sure it wasn't the Danish bloke?'

'He says someone came and took her after he'd hauled her out,' Johannes goes on calmly. 'Someone who came up out of the sea.'

Harvey's gaze alternates between Johannes and me and then, shaking his head, he peers at me over his coffee cup. 'Had you been at the bar cupboard, eh?'

'Did you see anything of the police chief and his sergeant?' I ask, ignoring the insinuation about my booze habits.

'Bjørkang? No. What do you mean? Are they out now?'

'They were supposed to pick me up from the lighthouse last night, but never turned up.'

'By the way, we need to let them know that the whole jetty broke free last night and is now drifting somewhere out in the fjord,' Johannes interjects. 'It would wreck your boat, you know, if you drove straight into it.'

'Have you phoned?'

'There's no answer. Not at home or at the office.'

'Neither of them?' Harvey sits further forward on the sofa, his eyes more alert.

'No.'

Harvey picks up his mobile phone and selects a number from his contact list. 'Answering machine,' he says, disconnecting. 'Have you tried to call the Skjervøy office?'

I shake my head.

Keying in the police station number, Harvey starts to pace the room. After a while, he hangs up and comes back to sit on the sofa again. 'You didn't hear any boats last night?'

I shake my head. 'Nothing at all, apart from the sound of a trawler engine late yesterday afternoon. Why do you ask?'

'The local police chief on Skjervøy says that the boat has been berthed in the marina at Blekøyhamn for the past few weeks. He hasn't been in contact with Bjørkang, Arnt or Tromsø over the weekend and knows nothing about picking anyone up from the lighthouse last night.'

'I don't like this,' I say, draining my coffee cup in a single draught. 'Is it far to this marina?'

Harvey stands up. 'I'll drive.'

CHAPTER 23

Sheets of rain come lashing at Harvey's car as it speeds through puddles and slushy ditches, close packed on either side of the road. Harvey has to take his foot off the accelerator to avoid skidding on the slick surface.

'This is going to turn out badly,' he says, while the vehicle toils up a steep hill with trees swaying violently in the wind. 'Really badly. And I have to go back out to the farm and find some way of attaching the sinker weights in case they break adrift again.'

'Aren't you scared?'

'Fuck, yeah, I'm scared,' Harvey exclaims. 'But what can I do? If I lose those mussel poles, that'll be the end for next year.'

'Do you think Arnt and Bjørkang are out there? In the storm?' The waves crash over the pebbles on the shore, and the threads holding dry fish heads smack against the frame of a fish rack down below the road. Somewhere in the midst of the grey weather I can make out Blekholm and the roof of the keeper's house.

'No idea,' Harvey says as we crest the summit and continue along the road towards the centre of Blekøyvær and the marina. 'But even if they are, the ambulance is an all-weather boat.'

I notice my body is still aching from its exertions on the rocks the night before, and my anxiety about the events of the past twenty-four hours, the sense that this is just the start of something, grows hugely with every minute that passes.

'What does the boat look like?' I ask once we finally turn off on to a minor road leading down a steep incline to a woodcrafts shop and a grocery store.

'Bright yellow,' Harvey replies as the car rolls slowly past people stooped and scurrying, hopping over puddles and quagmires between the shops and vehicles in the car park. 'With a white crane on deck.'

We continue to another car park in front of the marina, where a boathouse with a sign proclaiming *Blekøyvær Coast Association* stands in front of a horseshoe-shaped stone pier.

'It's not here,' I say when he stops at the car park in front of the boathouse.

'I see that.' Harvey takes out his mobile phone. 'I'll call Skjervøy again.'

'Hi! The boat's not here in the marina.' He drums his fingers on the steering wheel. 'Wait, and I'll check.' He gives me a brief nod before disappearing out of the car, heading for the boathouse.

As soon as he is out in the rain, I find my own phone and call Anniken Moritzen.

'Where are you?' she asks when she eventually answers. It sounds as though she's just woken up, or maybe she too has been to see Ulf and wheedled some more medicine, more of those painkilling, sleep-inducing preparations to sustain the numbness and keep some distance between day and night.

'I'm at the marina in Blekøyvær.'

'I thought you were on your way home.'

'We're searching for the boat belonging to the police chief and his sergeant.'

'Why?'

'I found a dead woman in the sea.' Thrusting my hand into my jacket pocket, I open the packet of OxyNorm and pop out two capsules. 'At Rasmus's lighthouse.' I gulp them down. 'The local police chief and his sergeant were to pick me up from the lighthouse last night after I spoke to you, but they never turned up.'

'A woman,' she says hesitantly when I'm done talking. 'Are you sure?'

'Yes. She'd been in the water for some time, but even though it's not always easy to see, it's definitely not him, Anniken. It's not Rasmus.' I realise I'm regretting this phone call. It's far too early, and I haven't taken time to plan the conversation in advance.

'I have to see for myself.' Anniken's voice has lost its listlessness, and she is speaking faster now, pursuing the words as panic spreads through her body. 'Maybe you're mistaken, a mother—'

'I'm not mistaken.'

'But how can you know? After all, you just said yourself it isn't easy to see.'

'Don't come,' I interject as a black figure emerges from the boathouse, breaking into a run as he heads through the rain to the car. 'Not yet. There's nothing to find up here. I'll phone when I know anything further.'

'No, wait,' she whimpers desperately. 'I don't understand—'

I hang up. 'Bloody idiot,' I mutter to myself as I put my mobile back in my pocket. 'What the hell are you doing, Thorkild?'

Harvey jumps inside and slams the car door. 'No one can get hold of them. The ambulance boat has been here in the marina all week, but was gone this morning when the first person arrived.' He turns on the heater full blast, blowing on his fingers. 'The police chief on Skjervøy has talked to the police in Tromsø as well as to the rescue coordination centre.'

'What now?' I feel my phone vibrate in my pocket, but let it be.

'They're sending a boat up,' Harvey says. His eyes follow the wipers as they race back and forth over the windscreen. 'They'll probably find them soon. The ambulance boat is rock solid and its equipment tip-top. For all we know, they're just having engine trouble or they've gone to one of the islands farther out.'

I attempt to see past the wipers to the rain-soaked seascape in the distance. 'I just spoke to his mother,' I say. 'Rasmus's mother.'

'What did you say?'

'I put my foot in it.'

'In what way?'

'I gave her hope.'

Harvey leans over the steering wheel and looks at me. 'Well,' he begins, 'there's something I have to ask.'

'Ask away,' I murmur, without taking my eyes off the rain dance out there on the pier, above the boats and the surface of the ocean.

'What actually happened at the lighthouse last night? With this ... woman you claim you found?'

'She was floating in the sea: a young woman, maybe in her twenties. With no face, and one lower arm missing. Wearing a nightdress and a T-shirt on top. Barefoot. As if she was lying asleep in among the seaweed.'

'And then you say a man came up out of the sea and took her?'

'Yes.'

'Did you see who it was?'

'No.'

'Were you drunk?'

I turn to face him. His complexion is glistening with raindrops and his hair is sodden. 'No.'

'Fine,' Harvey roars with laughter, tapping a beat with his fingers on the dashboard. 'Just had to ask.'

'I had a kind of accident some time back,' I start to say, directing my gaze out into the rain once more. 'Sustained a brain injury, some damaged connections that no longer work as they should. Sometimes I see things that might not be there. Smell, or feel someone close to me even though I'm completely alone. You start to doubt yourself, your own senses, but this isn't one of those instances.' I speak the final sentence mostly to myself.

'Hey,' Harvey fleetingly raises one hand before gripping the wheel again. 'Don't think about it. I believe all that stuff, you know that. All the way.'

The boats, tied up in serried ranks alongside the pier, are rocking lazily from side to side, tugging at their ropes between the waves.

'Do you think it was one of them?' he asks.

'Who?'

'That it was either Bjørkang or Arnt you saw come out of the sea and take the body?'

'Maybe it was Rasmus.'

'Rasmus is dead,' Harvey answers.

'Sure? All we have is his boat that drifted ashore in the cove. No body, no crime scene, nothing.'

Harvey looks at me in surprise. 'OK, I'll give you that,' he says. 'But why would he hide somewhere out on the island like some sort of ghost?'

'Maybe he had something to do with the faceless woman?'

'And doesn't want anyone to know about it.' Dimples suddenly appear in his cheeks. 'Aha, I see what you're doing. This is some real detective shit you're talking about.'

'I need somewhere to stay.' I take out another OxyNorm and pop it into my mouth, in the hope that it will tip the balance in my battle against these stomach pains.

'So you're planning on staying?'

'Looks as if I have no choice.'

———

Harvey swings out of the car park behind the boathouse and sets off on the return drive to the main road. My mouth is completely dry, and I can't produce enough saliva to swallow the capsule that is now stuck to my tongue. I have to dislodge it with the aid of my front teeth before rolling it between my molars and chewing hard. It tastes intensely sour, and I lose no time swallowing the contents before spitting the remainder of the gelatine capsule onto my hand and slipping it into my jacket pocket.

'Do you know of a guest house hereabouts, or somewhere I can rent a room for a few days?' I ask, cleaning my teeth with my tongue. 'It'll be too far to commute between here and Tromsø, and I'm not keen on having to sleep in the hire car until I wind things up here.'

Harvey stops at the main crossroads and turns to face me again. 'I might have a solution that would suit us both,' he says. 'We'll pick up your hire car first, then you can just follow me.'

Chapter 24

I park the hire car beside Harvey's truck and head out across the car park in front of the Skjellviktun Residential Centre's main entrance. The downpour is over but the wind is stronger here at the summit of the hill above the bay.

'Are you kidding?'

'No,' Harvey says, laughing, nodding towards the building directly opposite the centre. 'I bought the place six or seven years ago – it was originally built as a shower block for German soldiers during the war. The local council didn't use it, so I bought it, renovated it and divided it into three units that I rent back to the council care services. I've also got a similar project in Tromsø, where I own six bedsits that I rent out.'

'Enterprising,' I mutter, leaning against the boot of the hire car.

'Every little helps. All the same, one of the apartments is empty at present and you can rent it for, well, let's say three hundred and fifty kroner a day?'

'OK, then,' I say, glancing suspiciously at the building in front of me. 'I guess it'll be better than sleeping in the car.'

'Brilliant.' Harvey holds out his hand. 'You'll pay in advance, won't you? By the way, it'll be a bit difficult to give you a receipt – hope you don't mind that?'

'That's fine,' I answer, handing over the money just as Harvey's wife appears on the scene. She is dressed in an unbuttoned brown winter coat with mottled grey fur collar, white hat, jeans and matching heavy shoes trimmed with the same grey fur.

'Hi,' she says, wrapping her hands around her husband's neck. 'You got back safely from the farm. Thank goodness for that.'

Harvey nods. 'Yes – I had to drive through the valley.'

'Have you just arrived?'

'No, I've been to Blekøyvær with Thorkild.'

'Oh?'

'Bjørkang and Arnt are missing.'

'Missing?' She lets go of Harvey and takes a step back. 'What do you mean?'

'They went out in the ambulance boat last night to collect Thorkild from the lighthouse. He says they never turned up.'

As Merethe pulls the fur collar more snugly against her cheeks, her neck, mouth and nose disappear into the fake fur. 'Do you want me to phone Mari?'

'Not yet,' Harvey replies. 'We'll almost certainly have more information shortly. Wait at least until I come back.'

'What?' Merethe grips her husband's arm. 'Are you going out again?'

'I must, but I'll just stay inside the cove and come home again as soon as I've attached the mussel poles tightly enough. In the meantime, you can help Thorkild here, can't you? He's renting Andor and Josefine's old apartment for a few days.'

'I don't know if we can just rent out apartments to people. After all, the Department of Health and Social Services already has a contract with us.'

'Relax, honey – we're only talking about a couple of days. Anyway, there are no new residents due, according to what I've been told.'

Shaking her head, Merethe lets go of her husband's arm and takes a step in my direction. 'OK, then, Thorkild, come with me and I'll show you to your … accommodation.' She turns back to Harvey and says: 'Be careful.'

'Always,' Harvey assures her. 'We'll talk later, Thorkild.' He waves us goodbye and drives off.

'What does an occupational therapist do?' I ask as we cross the car park.

'I organise leisure activities for the residents here through the volunteer centre, such as pétanque, for example, therapy sessions and excursions. We arrange a religious service every third Sunday with the local vicar in the residents' lounge. I also do some extra shifts here at the centre when required. It's a good way to spend time with my mother since my father died and she was left on her own, and at the same time earn a few extra kroner.'

'So your mother lives here too?'

'Yes, she has a room in the dementia unit.'

Merethe produces a key when we reach the apartment at the far end and opens the door. 'Here,' she says, handing me the key before we step inside. 'The man who used to live here died while they were on a bus trip to Sweden with us. Heart attack.'

'Where's his wife?'

'She died that same night. *Of a broken heart*,' she adds in an American accent. 'It is possible, in fact,' she continues when she catches my eye. 'Sad, don't you think?'

'Yes, very sad,' I agree.

'I always get a lump in my throat, thinking of that sort of thing. Oh, and the funeral is on Wednesday. Their daughter lives in the south of Sweden and didn't want to come to clear out the apartment until after the funeral. We've stashed their belongings in the storeroom and the bedroom. I suggest you sleep on the sofa while you're here.'

I immediately regret removing my shoes. The floor is icy, like the whole apartment. No heating, and no light in the gloomy living room furnished with an old sofa, chair, sideboard, bookcase and TV table. The rugs are rolled up and stowed in one corner beside a box marked 'Books' and another labelled 'Pictures & Misc'.

'Harvey says you're some sort of medium?' I venture as Merethe heads for the fuse box. She is no more than about five foot two and has to stretch on tiptoe to reach the fuses.

'I'm what you call clairvoyant, but also do some healing touch and crystal therapy here at the centre from time to time.'

'And you're soon going to be on TV? On *Spiritual Search*?'

'Powers,' she says, laughing. '*Spiritual Powers*.' Crossing to the window, she opens the curtains and switches on the panel heater. Soon the entire apartment reeks of burning dust. 'We'll start filming after Christmas. Exciting, isn't it?'

I nod and sit down on the sofa beside her. Merethe wears big rings with colourful stones on most of her fingers.

'We humans can communicate with one another through healing energies.' She runs her fingertips along her tight jeans. 'Energies that we all carry within us and that cross the limits of the physical, spiritual and psychological worlds.'

'You speak to them?'

'Yes, all the time.'

'What … what do they look like?'

'What do you mean?' Her nails make a rasping noise as she runs them back and forth across her jeans.

'Whether you—'

'You're asking whether I see dead people, ghastly corpses and stuff like that?'

I nod slowly.

'Good heavens, no,' she exclaims with a burst of laughter and lays her hand gently on my thigh. 'Who on earth would manage to live like that?'

'Yes, true enough,' I mumble.

'All life is surrounded by its own energy field that some of us can see and sometimes also communicate with.' Blinking, she pats my leg with the flat of her hand. 'Our bodies are electric, Aske. Didn't you know that?'

'No.'

Merethe places her hands on the table, palms facing up. The stones rattle as she raps her knuckles on the table surface. 'Do you want to tell me about her?'

'Wh … who?' I croak, noticing that this has jolted me out of the abyss of thought into which I have allowed myself to sink.

'Take it easy.' Merethe leans in closer and takes my hands in hers. They are warm, smooth; even the metal of the rings on her fingers is warm against my skin. 'I felt her the first time we met. A soft cloak that hangs about you, wraps itself around you. I can feel it now too.' She closes her eyes and uses her thumbs to caress the backs of my hands. 'Were you together?'

'No.' I automatically pull my hands away from her touch. 'I hardly knew her at all.'

Merethe opens her eyes once more. 'But you're connected?'

I refrain from answering, and slump back on the sofa, thrusting my hands into my jacket pocket where I keep the blister pack of OxyNorm.

'I think she's angry with me,' I finally tell her.

'Angry?' Merethe looks quizzically at me. 'Why would she be angry?'

'She doesn't want to come any more.' I note how my voice trembles and at the same time how hard it is to breathe while I talk. 'No matter how much I try, she doesn't want to come.'

'So you know she's there?'

'Of course. She did come back, you see.'

'Back?' Merethe cocks her head slightly. 'What do you mean?'

'I was driving, she died, then she came back,' I say, as if it was the most natural thing in the world. 'But now she doesn't want to come any more. I don't know what I could have done, whether it's this place or whether there's something wrong with my pills. I've already thrown away some that didn't work.'

Merethe stretches out her open hands towards me again, making a sign that I should do the same. I let go of the blister pack in my jacket pocket and do as she wishes. 'I need to speak to her,' I whisper. 'I don't know how long I'll have the strength to keep up this kind of thing.'

'Thorkild,' Merethe begins. 'As a clairvoyant, I pick up information in the form of feelings, images, smells and symbols that I can try to interpret and then impart to the person I'm working with. What you're talking about, being a spiritual medium, is something completely different.

That relies on lending part of your own self to the spiritual world. It's a very painful business, and I'm reluctant to hold séances like that because of the stress they cause.'

'Please,' I whisper, clutching her hands. 'I must know. I can't go on like this.'

Merethe stares earnestly at me for a long time before she looks away. 'OK,' she says, patting my leg carefully with one hand. 'I'll give it some thought, Thorkild.' She stands up. 'Just give me some time.'

I remain seated, motionless, for several minutes after she has gone. In the end I walk over to the sink, fill a glass with water and drink it down with two OxyNorm tablets before returning to the sofa. I try to think of Frei, but don't succeed.

CHAPTER 25

At five to two, I hear a knock on the door, and rise from the sofa where I've been dozing since Merethe left.

When I open the door, the woman outside holds out an unnaturally tanned hand and we make eye contact. 'Siv,' she introduces herself. 'I'm a nurse here at the centre.' Dressed in a lilac uniform, she has clogs on her feet and blonde, shoulder-length hair.

'Pleased to meet you.' I return her handshake.

Siv, small and slight with rough hands and bitten nails, can hardly be older than forty-something, but nevertheless her skin is paper-thin and wrinkled as if she is a chain smoker with her own sunbed.

'I've to show you the canteen.'

'OK.' I grab my jacket from the hook in the hallway, put on my shoes and close the door behind me.

'Skjellviktun Centre has three sections,' Siv starts to explain as we hasten along the short path between the two buildings. She talks in a detached and mechanical voice, as if I'm a relative with a senile father waiting outside in the back seat of my car, or maybe a new patient.

The main building comprises a rectangular timber construction with three storeys jutting out, one facing the road, and two at the rear at right angles to the main structure. 'We have a total of thirty-nine single rooms, consisting of long- and short-term rehabilitation places and respite places. The two floors at the back are the nursing home sections, but one is a dementia unit and one a somatic nursing unit, mainly for terminal care.'

The canteen, a rectangular room with twelve dining tables placed against the walls and four in the middle of the room, overlooks the rear of the building so that you can see the windows of the two storeys behind it with an emergency exit and metal staircase.

Eleven residents sit inside, spread around four tables. Two couples are seated at tables by the entrance, while four elderly men are huddled in a corner by the window. Suddenly I catch sight of Johannes in the company of two old ladies at a table by the wall.

'This is Thorkild Aske,' Siv announces in a loud voice, with military precision, glaring across at the noisy group. 'He'll be staying in Andor and Josefine's apartment for a few days.'

'But where will Andor and Josefine sleep, then?' a squeaky voice pipes up at Johannes's table. The frail woman is slumped in a wheelchair and her dentures dance around in her mouth when she speaks.

'They're *dead*,' the other woman reminds her. Plump, she is dressed in a flowery short-sleeved summer dress. Her silvery grey, tousled curls are askew on one side, while the other side is flat, as if she has just got out of bed.

'Is he Agnes's son?' the squeaky voice in the wheelchair enquires.

'No, stupid,' the other woman answers. 'Agnes's son is much better-looking.'

'He may be pretty, but he's a real idiot,' the woman in the wheelchair cackles, followed by a condescending nod in my direction. 'Poor soul. Are you lost?'

'Thorkild Aske,' I say as loudly as my voice allows, making a dramatic bow to the two women as I approach their table.

'This is Bernadotte.' Siv, who has followed me, launches into introductions. 'And the beautiful lady in the wheelchair is Oline.' I shake hands with each of them and give Johannes a nod.

'What are you doing here?' I ask, taking a seat once Siv has gone off to one of the other tables.

'I visit my sister in the dementia unit a few days a week, and then I like to have dinner here too,' Johannes answers. 'Besides, today they're holding a memorial service for Andor and Josefine in the residents' lounge afterwards.'

'Good gracious, look at how thin you've become.' Oline in the wheelchair leans towards me and runs a cold hand over my arm. 'Doesn't Agnes have any food for you in the house?' She goes on stroking my hand while she adjusts her dentures, gazing at me all the while with a sad expression. 'Her little boy,' she says, smiling, and pats my hand, before rooting around in the handbag on her lap. 'Here,' she whispers, as she places a coin in my hand. 'Then you can buy an ice cream later.'

'Thanks,' I say, about to explain that I don't need ten kroner, and that I'm big enough to buy my own ice cream, when Johannes interjects: 'The food here isn't what it once was.' Just then, a young man emerges from the kitchen with a food trolley and begins to serve the people seated at the tables.

'Oh?' I ask, feigning curiosity as I clutch the coin that Oline has given me. Deep down, I can feel how the aroma of food has triggered a spasm of pain in my belly.

'The chef.' Bernadotte leans across the table to Oline and me on the opposite side. 'People like that, you know.' She shakes her head as the young man approaches.

'This is Zin. He's from Burma,' Oline says at the very moment the man arrives at our table to serve the food. 'No, no,' she adds eagerly, pointing at me as Zin is about to put the plate in front of her. 'Give it to this lad. I don't want any of that stuff.'

'Apologies.' I stop Zin, on the point of placing a dinner plate in front of me, and run my hand lightly over my stomach. 'I've got indigestion.'

'They don't know anything about fish, you see,' Bernadotte perseveres. 'That's the problem. It's a completely different culture, you understand.'

'Mhmm.' I nod in embarrassment as Zin continues to serve the food, not appearing to pay any attention to the conversation.

Oline opens the lid of the meal that Zin is finally permitted to set down in front of her. Pulling a face, she replaces the lid and turns to face Zin with a hearty smile. 'When is Sofia coming back to the kitchen?' she asks in a tremulous voice.

'Her leave ends in the middle of January,' Zin answers in broken Norwegian.

'Tell her we miss her.'

'Shh, there he is,' Bernadotte interrupts, nudging Oline's hand as a short, stout man with black wiry hair appears at the kitchen door. 'That bloody Lapp from Lakselv.'

'What did you and Harvey find out?' Johannes asks, leaning across the table while Zin disappears into the kitchen again. 'Did you see anything of Bjørkang and his sergeant at the marina?'

I shake my head. 'They picked up the boat some time last night. Since then, nobody has heard a peep from them. They're sending a lifeboat from Tromsø. Possibly also a helicopter in the course of the day.'

Johannes looks at me as he shoves a forkful of potato into his mouth. Chewing quickly, he swallows it down with a gulp of water. 'Not good,' he mutters, shovelling up more potato and stuffing it into his gob. 'Not good at all.'

The next minute, Zin returns with a packet of prunes and hands them to me with a broad smile. 'For your stomach.' He bows briefly and vanishes into the kitchen again before I have had a chance to thank him.

'They'll help,' Olin says, eating her dessert with a serious expression. 'Prunes speed things up, you'll see.'

'Yes – I'll have to go back to the apartment and give them a try,' I answer, and rise to take my leave.

CHAPTER 26

The prunes didn't work. I sit on the sofa, listening to the hiss from the radio, when my mobile phone rings. A man tells me the call is from Police Headquarters in Tromsø and requests me to report there the following day at eleven o'clock sharp. I ask if they've heard from the local police chief or made contact with the boat, but the man merely reiterates that I have a meeting with them tomorrow at eleven o'clock, and it's vital I keep the appointment.

Then he cuts me off.

I decide to have a bath to see if the hot water will help. In the bathroom I run water into the tub and take off my clothes. I sit on the toilet-seat lid as I watch the water gushing out of the tap and the steam rise. I fleetingly recall the smell of the showers in Stavanger Prison and dwell on those minutes suspended at the end of the skipping rope; I think of the silence and relive the pain. My thoughts turn to Frei and the barriers she has to cross in order to return.

As the water sloshes in the bathtub, I spring up from the toilet seat and peer expectantly over the edge just as my mobile sounds in the living room. I wheel round and dash across the chilly floor to take the call.

'It's me,' Liz pants breathlessly. 'I'm here.'

I open the curtains and look out at the headlights of a car in the car park with its engine idling. 'What are you doing here?'

'I saw your hire car.' Liz coughs loudly into the receiver. 'Arvid has gone on a trip to the mountains with some friends. I thought I would come and visit you.'

'How did you know where I was?'

'I drove around and—'

'You could just have phoned.'

'Yes, but I wanted it to be a surprise. I've baked some cupcakes. I don't know what they'll taste like, but I thought we could—'

'The apartment at the far end right in front of you,' I cut in. 'I'm in the bath.'

Opening the door, I return to the bathroom and turn off the tap. The bath is three-quarters full. Cautiously, I climb in and lean back so that only my knees and head jut out from the surface of the water.

'Hello?' I hear the rustle of shopping bags in the hallway and the front door slams shut. 'Thorkild, are you there?' Liz prattles nineteen to the dozen. I can hear her take off her jacket and kick off her shoes in one and the same energetic, uncoordinated movement.

My sister's emotional pendulum swings vigorously between two extremes: one a chaotic orgy of carbohydrates and self-reproach, the other blind faith in cosmic goodness and justice whose like happens only in fairy tales and Disney films. She also possesses a unique talent for bouncing back after setbacks, always a few kilos heavier than before, but she bounces back, camouflages her bruises, covers up the mental abuse and lets goodness stream in again. Fills her senses with desperate fantasies about the perfection of fellow human beings, whether that be husband, friend or brother.

'Hey, are you there?' she babbles at the doorway. 'You're having a bath, then?'

'Washing away my sins,' I say, running my rough finger-tips over my face and hair.

'You always had to have a flannel over your eyes when you were little,' she says.

'Didn't everyone do that?'

'No,' she giggles. 'I didn't. We weren't even allowed to rinse behind your ears with the shower spray because you were scared the water would go inside your head.'

'Your Honour,' I mumble, half-smiling. 'The witness is lying.'

'You thought you might get – oh, oh' – she is shaking with uncontrolled laughter now – 'water on the brain! Oh my goodness,' she adds, 'heaven only knows where you got that idea from.'

'Lies and more lies.'

'Don't say that, Thorkild.' Liz tells me from behind the closed door. 'You know it's true.'

'Well, off you go,' I say, 'and let an old man finish bathing.'

Still laughing, Liz moves from the bathroom door with her shopping bags and I close my eyes, immersing myself in the hot bathwater. As soon as I'm submerged, I open them again and look up at the opaque surface as the soapy water stings my eyes. I stay stretched out like that until I can no longer hold my breath. Then I clamber out of the tub and get dressed.

By the time I come back to the living room, Liz has set the table with paper plates, cups, napkins and a couple of big cake boxes filled with brownies and cupcakes, alongside a pot of freshly brewed coffee.

'Come and sit down,' she says. 'I hope you like them – I've made them all from scratch, though I'm not sure if the brownies aren't maybe a bit too hard, but—'

'I'm not hungry,' I tell her.

'Oh.' Liz gives me a sorrowful look.

I flop down beside her on the sofa and pour myself a cup of coffee. 'I found a dead woman in the sea last night,' I say, reclining with the cup in my hand. 'She didn't even have a face.'

'What?' Liz drops her half-eaten cupcake – topped with pink frosting and sprinkled with glitter balls – on to her plate, where it tips over.

'I phoned the local police chief. He and his sergeant were to come and pick me up, but they never appeared.'

Liz looks at me in silence.

'And now they're missing. I've to report to the police in Tromsø tomorrow.'

'The police?' She hesitates before continuing: 'Do they know ... who you are and what happened in Stavanger?'

I shrug as I drink my coffee. 'Almost certainly. Anyway, things are different now. I don't have any choice.'

Liz grabs the half-eaten cupcake and stuffs it all into her mouth. 'Last time, they took your job away,' she begins, before chewing as if her life depended on it. 'And sent you to jail.' She swallows and chews, chews and swallows, still struggling to complete her line of reasoning. 'You who ... who—'

'They couldn't do anything else,' I break in. 'What happened in Stavanger was my own fault.'

'How can you say that?' Liz puts two brownies on her plate and takes a third in her hand. 'That girl and her boyfriend spoiled everything for you. Everything.'

'No, Liz, they didn't,' I insist. 'You don't know what happened. What I did.'

'Because you won't tell me, Thorkild.' Her voice is softer now. 'Not even what happened while you were inside, about what you did. To yourself.'

'There's nothing to talk about.'

'But we can, if you want. Arvid will be away until tomorrow. I can stay here with you. Maybe we can just stay here, you and me, and have a chat.'

'About what?' I ask her, showing her that repulsive crooked smile of mine. 'Shall we start with me and my brain injury, or will we take you and your bruises first?'

I stand at the window watching Liz scuttle tearfully across the car park with her shopping bags. Once again I'm the one who has gone too far, who has found it easy to side with everyone who pulls her down, instead of being the brother she dreams I might be.

Outside, it is dark, and bare branches hang like the limbs of outsize insects above the car roofs. In the distance I see a light flash and glint in the coal-black sky above the bay and lighthouse. Immediately afterwards, the light breaks through completely and spreads across the whole island while a heavy throbbing emerges, forcing its way through the clouds.

As the light dances over the rocks and surface of the sea, the characteristic thudding follows it. The searchlight sways slightly from side to side as the Sea King helicopter heads gradually northward, seeking the boat with the missing police chief and his sergeant.

'I've made up my mind,' I whisper as I stand there with my eye on the metal bird, listening to the thumping wing beats. 'If you won't come to me, then I'll come to you.' The helicopter searchlight is disappearing out of sight and the noise is fading with it. 'I just have to get ready first.'

CHAPTER 27

My second to last day with Frei, Stavanger, 25 October 2011

Frei phoned the day after the dance class at the Sølvberg Centre. I was in a meeting with the deputy and assistant police chiefs, and we were working through a list of documents in connection with the investigation into the accused police officer.

'What are you doing?'

'I'm in a meeting.'

'When will you be finished?'

I looked at the clock before turning away from the table. 'What's this about?'

'Are you coming out with me today?'

I could see the assistant police chief watching me intently, while riffling through the pages of a duty roster. The deputy chief was tapping vaguely on his mobile phone, as he had done throughout the meeting. 'Well,' I demurred, 'all right then.'

'Brilliant. Pick me up when you're ready?'

'Where?'

'At Uncle Arne's.'

'OK.'

I hung up and gave the assistant chief a half-hearted smile. He nodded wryly before gathering up his belongings and dashing after the deputy chief, who had already left the conference room without a word.

Frei was sitting on the steps outside the villa in the Paradis district of Storhaug, leafing through a Thai takeaway menu, when I arrived. She stuffed the menu back into the mailbox and we set off together towards the city centre.

'What have you been up to today?' she asked as we crossed from Frue Terrasse to Hjelmelandsgata.

'Meetings, paperwork and planning for an interview I'm conducting tomorrow,' I answered. We were walking at arm's length. The sky was cloudless and the weather warm with almost no breeze. In a grove of trees on the upper side of Hjelmelandsgata near Admiral Cruys gate, we were met by a flurry of small birds flitting in and out between bare branches.

'An interview?' Frei was wearing a loose white top with rolled-up sleeves and tight, low-rise grey jeans, a shoulder bag and white trainers. I was still dressed in my work clothes, which usually comprised a plain, neutral-coloured shirt with sleeves turned up, no tie and dark suit trousers, an outfit I normally adopted even in the cold depths of winter because I had felt too old for jeans ever since my mid-twenties. I also preferred to buy a whole lot of the same thing when it came to clothes, with only certain colour variations between each garment.

'Don't you call it "interrogation" any longer?'

'Yes, I suppose so, but it's a handy Anglicism. Some of us prefer to call the sessions, especially those involving police personnel, conversations or interviews. It creates a certain distance between what many of these officers themselves do in their own work, and what awaits them with us in Internal Affairs,' I replied, smiling.

'And it works?'

'If only you knew.'

Frei looked at me. Her gaze was open, inquisitive, but contained something else as well, some kind of expectancy around the edges: 'So who are you holding a *conversation* with tomorrow, then?'

'A police officer at Police Headquarters here in Stavanger.'

'What's the case about?'

'Gross negligence in the line of duty.'

'And that means?'

'Don't you study law?'

Frei came to a halt. 'Do you plan the questions you are going to ask?'

'Absolutely. It's important to plan the session in advance, to single out the scenario and identify the tactically correct questions and follow a chronology in the order of questioning.'

'How can you know what you're going to ask in advance?'

'In an interview of this type the suspect has already provided a statement. It is the evidence, witness statements and assertions that the accused has introduced, as well as his personal statement, that form the basis for the interview I'm now going to conduct. I already have a good overview of the case and the people involved. The key thing is actually to hit on the tactically correct moment for this conversation.'

'What do you mean?'

'Well, in every investigation it's usual for the accused, or witnesses to the incident, to be questioned first, and in addition the accused himself has given his explanation of the incident or events, isn't that so?'

Frei nodded, and I went on: 'We collect the necessary information about what took place first of all, in order to devise a credible sequence of events, a scenario. Only then is it correct from a tactical point of view to conduct an interview with the suspect, so that we know enough to ensure the information that emerges can be controlled and possibly refuted in the process.'

'Who reported him?'

'Did I say that he'd been reported?' I pulled a smile and shook my head.

'So you have already spoken to the person in question who' – hesitating, she gave a lopsided smile before she continued – 'might possibly have reported the officer?'

'Of course. It's crucial to clarify as early as possible whether this is a formal complaint, in other words an internal disciplinary matter for the police station, or whether a crime has been committed. Most cases we receive have been wrongly directed to Internal Affairs and more often than not concern simple formal complaints or disciplinary matters where no crime has taken place. We usually sift these out fairly quickly without any interview required.'

'But this case doesn't fall into that category?'

'No. The person I'm going to interview tomorrow is also involved in another case we looked into earlier but was dropped, so it's possible that the indictment will be altered eventually. We'll see.'

We walked on towards the city centre through a narrow street where a number of small birds had gathered in the trees on either side.

'Altered to what, then?'

'You know I can't answer that,' I said, smiling to myself. The stroll across this peninsula, populated by timber and stone houses, to the accompaniment of bird chatter and the rustle of autumn leaves, was doing me the world of good.

I felt like a different person, not Thorkild Aske with all his defects, rules, facts and places he had to be, people he had to meet and characterise according to standard criteria, but merely this nameless individual who was moved by the sound of footfalls on the tarmac, the scent of autumn and the woman by his side.

'Is he in trouble?'

All of a sudden, Frei stopped again and looked at me, this time for longer than before.

'What do you mean?' We were standing in the shade of a massive tree, its trunk stained with green moss.

'Nothing.' Frei blinked rapidly before taking my arm and pulling me onward.

I came to a halt and held her back. 'What is it? Do you know this officer?'

Frei was reluctant to look at me. 'No, it's nothing,' she answered quickly. This time her eyes merely slid swiftly past mine. She leant lightly against me with her hands between us, and her face turned to the ground, so that her hair brushed up against my chin. 'Relax. I'm just curious, that's all.'

My head drifted slowly forward. I noticed that I was breathing through my nose, soundlessly and mechanically, as if struggling to remain calm and not move a muscle for fear that the tiniest motion would shatter the image in the water.

Without warning, Frei looked up at me, and before I had time to think I leaned all the way forward. My lips only just touched her top lip and the tip of her nose before she abruptly drew back. She retreated to the stone wall that stretched out on this side of the street. 'What are you doing?' Her eyes were wide open and she used her hands to hug her body as though in self-protection.

'I ... I'm sorry.' I took a step back, as if to give even more space to the distance between us that had suddenly opened out. 'My God, I thought, I—'

'No,' Frei gasped, shaking her head in desperation. 'We're not ... You and I can never ...'

Without completing the sentence, she simply turned on her heel. She broke into a run along the street and didn't look back.

I was left standing there, watching her leave, until she was gone. 'You're lying,' I whispered through gritted teeth. 'Time after time.'

In the end I turned tail and walked back the same way we had come, before heading from Storhaugveien in the direction of the bus station. My face was burning, a blaze of anger and pain that wove its way through me from the inside out, as I hurried off down the city streets towards my hotel.

I knew that I had gone too far. This was not part of the game. I was on the verge of losing control of the situation.

MONDAY

CHAPTER 28

I don't reach Tromsø until one o'clock. The ferry from Olderdalen to Lyngseidet had been cancelled, and I had to drive round. My entire body is aching after the long drive on rugged, bumpy roads.

The tide is all the way up to the edge of the quay, and as I turn in to Grønnegata, heading to the city centre, I see that the road is flooded. On both sides, workers from the Highway Authority watch as excavators pile snow and slush on waiting lorries.

A chill mix of snow and rain pummels my face and hair as I open the car door. Inside Police Headquarters I cross to the reception counter and announce my arrival. A stout woman in her mid-twenties asks me to wait, and I take a seat on an uncomfortable bench between two flower troughs filled with plastic plants and clay pebbles.

Five or six minutes later, a man emerges from a door and asks if I am Thorkild Aske. I nod and stand up when he beckons me. He disappears through the door long before I reach him and I find him waiting outside another half-open door at the end of a long, narrow corridor with pictures of former police chiefs on the walls.

'In here,' says the police officer, with a mug in one hand and a document folder tucked underneath his arm. He is holding the door open. 'We've been waiting for you.'

The room is bare, with fibreglass wallpaper on the walls and linoleum on the floor. All it contains is a rectangular table with two chairs on one side and a hard, uncomfortable chair on the other. I could have told them that nowadays we were supposed to sit as equals during these sessions, using similar chairs with a small table in front of them, rather than between us. At least the prevailing witness psychology insists that is the optimum stimulus for communication.

'OK, then.' The officer is tall and athletic, with bright blue eyes and a narrow mouth. 'Sit down.' He points at the hard, uncomfortable chair.

'Thanks.' I stop just in front of the table. 'Tell me, though, can't I take one of those other chairs? This one looks terribly uncomfortable.'

'That chair is perfectly OK, it—' he starts to say.

'Brilliant! Thanks very much,' I say breezily, exchanging chairs and making space on my side of the table. 'So, what are we to talk about today?'

'Just a moment,' he mumbles, keeping a lookout along the corridor. No one seems to be coming to his rescue, and in the end he closes the door before taking a seat on the remaining cushioned chair.

'Well, then,' he begins, opening his folder. 'We can make a start by filling this out,' he says. 'While we wait.'

'Terrific.' At once I notice how my unwillingness to participate in these surroundings, now that I am no longer in the driving seat, tugs and tears at me under my skin. That was also how it had felt in the interviews following what happened to Frei. Ulf would have remarked that my own misplaced professional pride is the driving force behind this infantile compulsion to prolong and sabotage situations such as this. As for me, I simply can't come up with another way of tackling it.

'We'll start with your personal details,' the officer continues.

'Don't you have computers up here?'

'Yes, of course.'

'Excellent: Thorkild Aske, date of birth January sixth, 1971.'

'Where?'

'In Skuflavik, Iceland.'

'Norwegian citizenship?'

'Yes. The certificate is lying about somewhere if you require it.'

The officer waves his hand without glancing up from the paper. 'Unmarried?'

'Divorced.'

'Employment?'

'Job-seeker. Are you hiring, by the way?'

'I understand you previously worked for Police Internal Affairs in Bergen, where you were a lead interrogator?'

'Correct. But not anymore.'

'No, I know that.'

'Of course you do.'

After a moment's hesitation, the man goes on without meeting my eye. 'You were previously found guilty of involuntary manslaughter and served a prison sentence. Tell me about—'

'Can't we just say that I know that you know, and then we can move on to whatever this is all about?'

Finally the man looks up at me: 'What can you tell me about the events that took place at ...' He leafs through some papers in the folder: 'Blekholm Lighthouse.' He jots this down on the paper and then adds: 'From Saturday the twenty-fourth of October up to and including Sunday morning the twenty-fifth?'

'So you haven't found them yet?'

'No,' he answers curtly, again without a glance in my direction. 'Tell me, when exactly did you arrive at the lighthouse and when did you leave?'

'I was there from six a.m. on Saturday until five a.m. on Sunday. What about the boat? Have you located it?'

Another man enters the room. He is older, even older than me, with silver hair, and is slim with a straight, pointed, aquiline nose of aristocratic quality and appearance. This man stops for a moment or two in front of the hard, uncomfortable wooden chair before sitting down when neither of us shows any sign of willingness to change.

'I am Superintendent Martin Sverdrup.' He proffers his hand. He is from the north of Norway but uses the hybrid dialect that you sometimes hear politicians and other people from these parts switch to when they are on TV or in civilised company.

His handshake is firm and reassuring. 'Thorkild Aske,' I reply before slouching in my chair again.

Martin Sverdrup rubs his hands and shakes his shoulders as if we are simply three good friends gathered around the fireplace to discuss the weather and fishing. So it is now time to build a relationship through trivial small talk. A method intended to make the eventual transition to harder questions proceed more smoothly. In other words, they also teach the KREATIV technique to police officers in northern Norway.

'Coffee?' Martin Sverdrup opens with a jovial, friendly nod towards the door.

'Who, did you say?' I decide I'm not going to take the bait today.

'Coffee, tea ... would you like anything?'

'No thanks.'

'Soft drink? Or water?'

'OK, water.'

'We don't have any fancy bottled water, but the tap water here tastes very good.' Martin Sverdrup steals a glance at his colleague with the pen in his hand and a blank expression on his face before returning his attention to me. 'That will be acceptable, I expect?'

'No, forget it,' I say. 'I'd prefer a cup of camomile tea.' I look across at the policeman who is still just sitting there, pen in hand and staring straight ahead. 'Or juice. Do you have any orange juice?'

'Er, I don't know if ...' Martin Sverdrup looks quizzically at his colleague. 'Steinar? Do we?'

'Do we what?' The man lets the pen fall on the table as if he has just been roused from a deep daydream.

'Juice. Do you know if we have any?'

'Forget it,' I interject. 'Coffee, then. Black.'

'Great. Excellent.' Martin Sverdrup focuses on his officer again: 'Would you run and fetch some, then?'

Steinar churlishly pushes the folder and sheet of paper he was in the process of filling out across to his boss, adding the fountain pen as an afterthought, before getting to his feet and disappearing from the room.

'OK, Aske.' Martin Sverdrup skims the fields already completed. 'Back to you,' he says. 'You know why you are here?'

'I think so, yes.'

'Excellent.'

'Yes, excellent.'

'Terrific. Well, let me get straight to the point.'

'Yes, let's do that.'

'First of all, fill me in on the background to your arrival here and your reasons for being out at that lighthouse.'

'The parents of the boy who vanished from the lighthouse at Blekholm some time last weekend paid me to come up here and search for him. His mother bought the lighthouse for her son Rasmus last summer, and he was busy renovating it with the idea of turning it into some kind of activity hotel. When I got there, I contacted the local police chief and met him and his sergeant at the police station. Later, they drove me to Skjellvik, and from there I arranged a lift over to the lighthouse the following morning.'

'And when did you next speak to Bjørkang or his sergeant, Arnt Eriksen?'

'That same evening when I found a woman in the sea.'

'A woman, yes. What can you tell us about her?'

'Young, early twenties, about five foot two, dark mid-length hair. Wearing a nightdress and T-shirt.'

'Facial features?'

'None.'

'None?'

'Her face was gone, and so was one of her lower arms. She appeared to have been in the sea for some time. The body was in the early stages of hydrolysis and had already acquired the soap-like appearance that—'

'And then she disappeared?'

'Yes.'

'All of a sudden?'

166

'Someone came up out of the water and took her away.'

'Did you see what that person looked like?'

'He was wearing a diving suit.'

'Did you see or hear a boat?'

'No.'

'So he came directly up out of the water?'

'Straight up. Like a rocket.' My hand describes a soaring action above the table that separates us. 'Swish, swish.'

The corners of the superintendent's mouth drop a notch or two as he struggles to ignore the hand movement and swishing sounds. 'Do you have any idea who it was or what happened to the corpse?'

'I don't like to speculate.'

Martin Sverdrup turns to a different page. 'What did you do then?'

'I phoned Bjørkang as soon as I had pulled her ashore and asked them to come and pick us up from the lighthouse.'

'What time was that?'

I take out my mobile phone and locate the call. 'Half five in the afternoon, and I spoke to the sergeant a few minutes later.'

'And when was the next time you saw or spoke to either of them?'

'I didn't.'

'Sure?'

'As shooting.'

'Are you taking any kind of medication at present?'

'Loads.'

'OK.' Sverdrup uses a finger to scratch between the knuckles on his other hand. 'Sorry about that. Let's both take a breather before we continue. What do you say?'

'Fine,' I answer, my shoulders slumping. 'Have you found them?'

Martin Sverdrup shakes his head slowly.

'Nothing?'

He shakes his head again. This time even more slowly. 'We're searching the area where we think the ship was wrecked. It will be undertaken as a systematic search for a wreck using an echo sounder, and we calculate that the chances of finding them are pretty good. Sooner or later.'

'Preferably sooner, rather than later,' I interject.

'Do you have any idea why they would have brought diving equipment with them on the trip?'

'Sorry?'

'Diving equipment. Diving equipment that can't be accounted for has been removed from the boathouse in the marina. We believe they took it with them in the boat.'

Why on earth would Arnt or Bjørkang put on a diving suit and head out to the lighthouse just to pick up the body and leave me sitting there like a fool? And if it was one of them I saw, where are they now? It strikes me that if this is the scenario the police are now working on, that they did in fact come to the lighthouse while I was there, and subsequently disappeared, then the outcome might go hard if they don't turn up again soon. Extremely hard.

'Who goes diving in a storm?' I finally ask.

'Well, according to you that was exactly what someone did – the person who stole this woman's body you claim to have found.'

'Maybe I only dreamt it, then,' I respond crossly.

'Is that a possibility?' Sverdrup asks with some misgiving.

'No.'

'Sure?'

'What am I doing here, Martin?' I ask in annoyance, opening out my arms. 'What if you tell me instead what my status is in this case?'

'Witness,' the superintendent says hastily.

'For the meantime,' a voice suddenly makes itself heard from the doorway, and a male figure slips into the room, his eyes directed at me. Stalking across the room, he takes hold of the vacant chair, pulls it away from the table and sits astride it directly opposite me. My former boss in Internal Affairs gives a brief nod as he scrutinises me from head to toe. 'Tell me, Martin,' Gunnar Ore continues, his torso inclined towards the chair back with his arms dangling over the top. 'How do you actually interrogate a lead interrogator? Do you have any idea?'

Chapter 29

'How's it going, Gunnar?' I ask.

'Marvellous,' Gunnar Ore answers. His powerful, hairy forearms seem even more muscular and tanned since the last time we saw each other, back when he was still my boss in Internal Affairs. 'What about you?'

'Like a square peg in a round hole.'

'Oh, as good as that?'

'Yes, to tell the truth.'

We remain seated, mutely sizing each other up.

'What are you doing here?' I ask in the end. 'Are you still in Internal Affairs?'

'No, back in Delta.'

'The Emergency Squad?'

'Correct.' Gunnar Ore nods without taking his eyes off me for a single second.

'Minding the store, eh?'

His narrow lips resemble a straight line scored across his suntanned face. Gunnar is fifty-three, but nevertheless every single muscle in his body tenses when he moves: even his facial muscles contract and relax easily and elegantly beneath his skin when he clenches his teeth to make his lips even narrower. When he took up his post as the new head of the Western Section of Internal Affairs, from his previous job in the Emergency Squad, rumours were already flying around the place that he had taken part in the action at Torp two decades earlier, and that he was, to quote: *a sharp shooter*. Gunnar Ore counts on respect wherever he goes, and people found it natural to accord him that, whether they already knew him or not. It was simply something that coursed through his blood, and that you could not escape.

'Still on active duty,' he replies, smiling for the first time. 'Challenging assignments, just like the good old days. Need to feel the adrenalin pumping, you know, even though working out is a bit of a tougher grind these days. The years come creeping up on us all.'

'Why are you here, Gunnar?'

A hint of a smile crosses Gunnar Ore's face before the straight line returns: 'OK, then, I'll tell you, Thorkild. I got a phone call from a certain local police chief, Bjørkang, a few days ago. He wanted to know what I could tell him about one of my former minions in Internal Affairs, Thorkild Aske to be more specific, who of all things had announced his arrival in the hospitable north where he intended to search for a drowned man out at a lighthouse. I told the police chief in question that Thorkild Aske was a name I had hoped never to hear again. That this shameful individual who had chosen to betray all his friends in the force, as well as our reputation, me personally and everything we stand for, should turn up in a police investigation of all things, was really the very last thing I had expected.'

I can see his jaw muscles working as he speaks through gritted teeth. Gunnar Ore homes in on me: 'Have you any idea how bloody awkward it was to have to sit there defending a fucking DUI in my own team? Driving under the influence of drugs, involuntary manslaughter! How do you think that reflected on me, on us, on the rest of your team? Have you ever considered that?'

'Manipulation and information,' I comment after an uncomfortable hiatus with all three of us simply looking at one another and not saying a word.

'What the hell did you just say?' Gunnar Ore draws his chair closer. 'What did you say?'

'You wanted to know how to interrogate a lead interrogator. Well, all interrogation, when push comes to shove, is a question of two things: manipulation and information. Even when you interrogate someone who himself has experience and knowledge of the methods. The most important aspect is nevertheless what you put aside before you step inside the interview room.'

'And that is?' Martin Sverdrup lets his ballpoint pen dance between his fingers, as if here for a course on interpersonal police communications. Gunnar Ore, on the other hand, remains motionless, staring at me. He is so close that I am aware of the smell of aftershave and strawberry-flavour chewing gum.

'Yourself. The lead interrogator steps into the room as a blank sheet of paper, water in an aquarium, with no prejudices, no anger, rage or any other distracting factors. What you want is to make the subject aware of his or her own personal responsibility for the situation they find themselves in, while you check whether the conditions for criminal liability have been met. And then you have to build some sort of trust.'

'A kind of whore, in other words.' Gunnar Ore spits out the words, digging his fingers into the armrest before clenching his fist.

'If you like,' I answer, unruffled. 'Then the game commences: manipulation and information. Deception is an excellent place to start when dealing with someone who is already skilled in the game. Combine fact and fiction, create the impression that you know more than you actually do, confuse, make the interview subject uncertain.'

'Not the KREATIV technique, then?' Gunnar Ore has that half-smile on his face again.

'Ugh, no,' I answer, shaking my head. 'Not for these people, no. In these instances you need to resort to desperate measures. Psychological manipulation techniques exclusively, if you ask me. My suggestion would be to begin the session by probing into the subject's intimate zone, and to appear threatening. They probably won't anticipate that.'

'Like this?' Gunnar Ore pulls the chair a few centimetres closer to me, so close that our faces are almost touching.

'Brilliant. If I were you, I would now perhaps consider running a few rounds of the good cop/bad cop routine, with carrot and stick as incentives.' I nod at Sverdrup. 'You have manpower enough for that.'

'Nah,' Gunnar Ore says gruffly. 'What else do you suggest?'

'Ego pressure? A well-aimed attack on the subject's self-esteem in order to persuade him to justify himself and his actions by talking and explaining. The Reid Manoeuvre or lie detection is a dubious route to take, but worth a shot, especially in combination with one or more of the methods I've mentioned.'

'Is that all?' Gunnar Ore is whispering now, still between gritted teeth.

'No,' I continue calmly. 'You also have *Verschärfte Vernehmung.*'

173

'Yes,' Gunnar says, nodding gravely. 'Now we're speaking the same language.'

'What's that?' Sverdrup asks inquisitively.

'Advanced interrogation. The kind used by the Gestapo during the war, and by and large the same methods the Americans use today. Isn't that funny?'

'What else?' Now Gunnar Ore is drumming his fingers impatiently on the chair back. 'Is that all? Do you have any more tricks up your sleeve?'

'More? Well, you mustn't forget the administration of psychoactive drugs. But by then we've gone some distance into the realm of advanced interrogation techniques.'

'Rumour has it you do a good job of administering them to yourself these days.' I can hear the teeth grinding in his upper and lower jaws.

'But,' I plough on without paying any attention to him, 'there's always a possibility that the subject suffers from some form of personality disturbance, and then the rules of the game change very fast.'

'Why's that?' Martin Sverdrup is now clicking the ballpoint pen.

'Follow this closely,' Gunnar Ore says. 'He knows this inside out. He's on home territory now.'

'About seven per cent of the world's population suffer from psychopathy in some form or other. Half of all inmates in American prisons fall into this category and they in turn are behind eighty per cent of all violent crimes. Human beings without empathy, lacking personal insight, chronic liars with no remorse or impulse control.'

'Do you know what, Martin?' Gunnar Ore turns to face Sverdrup. 'In fact, we had someone like that in our section at one time. A real creep of a guy.' Then he turns back to me. 'There we have it, then. The psychopath, yes indeed.'

'What do you do then?' Sverdrup asks in a valiant effort to assume control of the conversation. Falling silent again, Gunnar Ore stares straight into my eyes as I press on: 'Well, it would be a total waste of time to attempt to make a psychopath acknowledge any kind of responsibility or remorse. They are unable to appreciate the physical, emotional or mental pain they inflict on others. Their behaviour is exclusively dictated by a narcissistic need to satisfy their own egos. The successful interrogation of a psychopath is only achieved when the interviewer is aware of this and makes no attempt to appeal to the person's sympathy, conscience or social attachments.'

'Here comes the entertaining part,' Gunnar Ore comments. 'Listen and learn.'

I nod, aware of where Gunnar wants to go with this conversation, but I can't come up with a way of stopping him. So I simply continue, steeling myself for what is to come: 'The lead interrogator should really pretend to be slightly impressed by the originality, ingenuity and strength the interview subject has shown in his exploits with questions such as – *How did you manage to murder someone by that method? So many? For such a long time without being caught?* And so on and so forth. In other words, you will only obtain a confession as long as the psychopath himself feels that it fulfils one or more of his own egotistic demands.'

'How do you know all this?' Sverdrup asks.

'Thorkild interviewed a number of these maniacs when he was in the States with this doctor of his. Dr Ohlenborg, wasn't that his name? Several of them were ... policemen?'

This is where Gunnar Ore has been steering the conversation from the very beginning. The place where he can encapsulate everything I am and have been, in this all-consuming truth that I am supposed to carry round my neck like a millstone for the rest of my life. This final proof of my treachery against him personally, against our team and against the whole of our police tribe. The stigma that shows what I will be like, now and for always, one of the type both of us once hated and despised more than anything else. 'So-called cops gone bad,' Gunnar Ore goes on. 'Police officers who change sides, who kill, rob, rape and destroy everything in their path.'

'Something like that, yes.'

'Just like you, isn't that right?'

'Go to hell, Gunnar!'

Gunnar Ore leaps up from his chair, snatches it up and hurls it across the floor before he grabs me by the throat and hauls me to my feet. 'Why can't you just go and die in a ditch?' he growls, before punching me right in the face with his free fist, making me lurch backwards, into the wall on the opposite side of the desk, and collapse on the floor.

'But what the ...' I hear the superintendent yell. I lift my head and see him using both hands to hold Gunnar by the arm, as if making a heroic effort to restrain a raging bull.

'That one was for the team.' Gunnar presses his fists together before tearing himself free from Martin Sverdrup's grip. He takes a couple of paces forward and kicks me in the stomach. The blow is so powerful that I throw up just as a stab of pain shoots through my diaphragm. A mixture of water, coffee and fresh blood trickles down my shirt collar and on to the floor. 'And that one's from me.'

'Bloody hell, man!' Martin Sverdrup grabs Gunnar's chest from behind. His ribcage is so broad that his arms only just reach round it. All the same, he manages by some remarkable feat to haul Gunnar back over to the table, where he sits down heavily in one of the chairs. Martin marches off to retrieve the chair that Gunnar has tossed aside and returns it to its rightful place before approaching me. 'Are you OK?' he asks, tentatively.

'Fine,' I say, launching a splodge of blood at the floor between us. 'No sweat. I'm fine.' I spit again and rest my head on the wall. Gunnar is seated at the table, his face buried in his hands, glowering at us through his fingers.

The superintendent takes a couple of steps towards Gunnar, raising his finger like a sword in front of him. 'Listen,' he begins, trembling with indignation. Suddenly he is speaking in pure northern dialect: 'I don't know what the two of you have going here, but this is absolutely unaccept-able. I intend to—'

'Get out,' Gunnar Ore whispers, his eyes glaring up through his fingertips at the superintendent, who stands frozen to the spot for a moment, with one finger held high in the air, as if to check the wind direction in the room. 'I want to talk to him on my own.'

'No, damn it, I'm the one who—'

'Out,' he says. This time with a darker undertone, a vibra-tion in his voice that forces Martin Sverdrup to hesitate and take a step back.

'It's OK,' I tell him, struggling unsuccessfully to get to my feet.

'Bloody hell,' the superintendent murmurs just as his colleague, the one who had gone out to fetch some coffee, finally puts his head round the door. He is about to say something, but Sverdrup forestalls him: 'Let's go, Steinar,' he says. 'These two lunatics need the room to themselves for a while.'

I wriggle on to my stomach, spitting more blood, before crawling across the floor to reach the chair beside the one Gunnar is using.

I catch hold of it and haul myself up. 'Well,' I say, coughing as I finally settle on the chair, finding some semblance of a seated position that I can live with. 'What are we going to talk about now, boss?'

Chapter 30

'You look fucking awful, Thorkild,' Gunnar Ore remarks as the door closes and the other two men leave the room.

'Why are you here?' I ask him.

'I could ask you the same thing.' Gunnar sprawls in the chair, rubbing the knuckles of one hand.

'Was it Ann-Mari who sent you?'

My ex-wife, Ann-Mari, and Gunnar got together while I was in America. It meant nothing to me. The divorce was already done and dusted and our marriage had been over for many years before we ultimately formalised the break-up.

'She doesn't know you're out.'

'Yes, she does,' I tell him before snatching one of the sheets of paper from the bundle of documents still lying on the table. I fold it and use it to wipe the blood from my face. 'She's started sending me more cuttings about children in the post. Got one just before I came north, of a boy and a girl in a clothing advert. You'll have to get her to pack that in.'

'But, bloody hell,' Gunnar says, once again covering his face with his fingers. 'She's just worn out these days,' he goes on. 'We've been trying for children for such a long time now, and—'

'Why not adopt?'

'Eh? Don't you think I've—'

'For God's sake, Gunnar.' I throw the blood-spattered sheet of paper down on the table that divides us. 'You know she can't have children, don't you?'

'What?' Gunnar Ore, fists clenched, is standing up again, but changes his mind halfway and flops back down on the seat. 'What the hell are you talking about?' he whispers, wearily.

'The doctors discovered these cysts in her ovaries. Nearly fifteen years ago. That's why she sends me the cuttings, to show what I deprived her of when she was still able, since I was the one who didn't want children. For fuck's sake, man,' I say, spitting, while Gunnar Ore goes on sitting on the chair in front of me, as if rooted to the spot. 'They had to remove everything for fear it would develop into cancer. Ovaries, uterus and lymph nodes, the whole shooting match, to prevent any spread. What do you think the scar on her stomach is?'

'She stabbed herself,' Gunnar suddenly whispers without meeting my eye. 'Down there, with a pair of scissors, the night after I came home and told her about you and that young girl in the car crash at Sandnes. A DUI, high on drugs, responsible for the death in a road accident of the girlfriend of the policeman whose conduct he's come to the city to investigate? Can you imagine the media circus all that might have led to if we hadn't acted fast enough?' Shaking his head, Gunnar slams his palms on the table again. 'All the work we had to do to keep the bloodhounds away from your shit. Both in our unit and at Stavanger Police Headquarters. It's almost a shame to have to say so, but it might have been better if you'd both—'

There is a knock at the door and immediately afterwards the police officer sticks his head round, bringing two steaming mugs of coffee. 'Er, here's that coffee,' he tells Gunnar, diffidently, as I sit up straight and take my mug.

'Not for me, thanks,' Gunnar replies, using his thumbs to rub his knuckles. 'Water, please.'

'We don't have bottled water, only water from the tap—'

'Same fucking thing, pal. A big glass, cold water. You get it.'

The man nods. 'Two seconds.'

'One – two!' Gunnar spurs him on without glancing up from his knuckles. The policeman's gaze lands on my face before shifting to Gunnar's knuckles. Then he turns on his heel and leaves.

'Do they really believe I'm involved in the disappearance of the local police chief and his sergeant?' I ask, blowing on my coffee.

'Of course not,' Gunnar answers. 'Actually, what the fuck do I know? I'm just passing through. I'm running a course for army officers up here next week. I just had to drop in to see if it really was you.' He laughs. 'And here you are, large as life.'

'And you too,' I reply. 'It's almost so cosy that we should have brought out the cocoa.'

The policeman returns with water and ice cubes in a tall jug that he sets down on the table. Gunnar takes a large swig before replacing his glass and rising from his chair. 'And by the way,' he says, pointing a rapier finger at my face, 'don't phone, OK?'

'That's a promise.'

Turning to go, he stops right beside the policeman standing to attention with his hands straight at his sides, as he stares at the table and the bloodstained papers extracted from his document folder. Gunnar turns to face me again. 'What's more, don't text, or call Ann-Mari either, don't visit my office in Grønland, don't send a postcard when you're on holiday, no emails, no Facebook messages, no embarrassing hellos from the other side of the street. Zilch. And keep away from our workmates in Bergen. Crawl back inside that car wreck, disappear down into the showers again; I don't give a flying fuck. Just vanish. You no longer exist, Thorkild Aske, is that understood?'

'Understood.'

'And another thing: vamoose from here as well. Out of this case, however things might turn out. You've no business here.'

'Soon,' I answer.

'Not soon. Now!'

'Goodbye, Gunnar.'

'Go to hell!'

Gunnar disappears through the door and two seconds later Sverdrup appears again, peering inside the room like a whipped dog. As soon as he has ascertained that the coast is clear, he slips inside and closes the door behind him. Both men resume their seats, and the policeman produces a fresh sheet of paper from the pile. He lifts it carefully over the scrunched, blood-spattered paper in the middle of the table, grabs his pen and starts to write. 'OK,' Sverdrup says, joining his hands tentatively in a reverent gesture intended to demonstrate calm and diligence. 'Well, that's over at last,' he comments, drumming his fingertips together. 'Let's take it all from the very beginning. In your own words.'

Once we are finished, and I have been asked to steer clear, but not to venture too far away, while they continue their search for the two missing police officers, I drive down to the shopping centre I had visited when I first arrived in the city. First of all I head for the pharmacy and buy some Duphalac before calling in at the perfume shop to buy the same bottle of perfume from the same shop assistant as last time.

Outside the centre, I see huge sheets of foam battering the Coastal Steamer quay. I get into the car and drive out of the city, northwards, through the darkness beneath the massive crags, past rotten fish racks and broken barn roofs that the wind has catapulted to the ground. I pass driftwood, poles and flotsam jutting from the mounds of seaweed at the edge of the foreshore, while gusts of wind drive tattered storm clouds in from the wide ocean.

Chapter 31

It is past five when I finally return to the residential centre at Skjellvik, and the wind has died down. The helicopter and lifeboat are gone, and the fjord is a pitching, blue-black expanse murmuring fitfully in the oppressive afternoon darkness. I take out my mobile phone and call Harvey.

'Yes,' Harvey answers in a relaxed voice. 'Mr Aske, is it not?'

'The very one,' I respond before Harvey roars with laughter. I can picture him in my mind's eye, raising a glass of fortified coffee to his kitchen window in a mock toast.

'They've found the boat,' he says more seriously.

'What? Are you sure?'

'Yes. Johannes just told me. He had heard it on his walkie-talkie. A pleasure craft found it this morning, washed ashore by the storm on Reinøya Island. Besides, I saw a police vessel out at the lighthouse earlier today. It looked as if they were there to undertake some investigations. They were all wearing white suits, and they left not long ago.'

'Scumbags,' I whisper.

'What is it?'

'Forget it. It's just that I bumped into an old acquaintance in Tromsø today.'

'Tromsø?'

'I've just been interviewed.'

'I see. And they didn't say anything?'

'No.'

'Why not?'

'Police tactics,' I tell him, leaning heavily on the side panel of the hire car. 'There's no point in telling a witness or possible suspect what they know until it's tactically advantageous.' Within myself, I go over the conversation at Police Headquarters with Gunnar Ore and Sverdrup a few hours earlier. It's no fucking coincidence that Gunnar has turned up here at the same time as two policemen have gone missing after arranging to meet me at that lighthouse. They've already begun to construct a scenario. A scenario that implicates me in Bjørkang and Arnt's disappearing act. 'Bloody hellfire, fuck it,' I groan, leaning forward as pain stabs through my stomach.

'Doesn't sound good, Thorkild,' Harvey comments.

'No, definitely not,' I moan, forcing my body up again. 'Tell me, where is Reinøya?'

'Farther south. Towards Tromsø. Talk to Johannes – he probably knows the fisherman who found the boat. I have to go back out to the farm before nightfall.'

'Bjørkang and Arnt took diving equipment with them when they went out that night.'

'How do you know that?'

'The police told me.'

'Doesn't sound like something Bjørkang would have agreed to,' Harvey says. 'He would never have gone out diving at that time – he's not very comfortable at sea even in good weather.'

'Could they have been involved in some criminal activity?'

'Why do you ask that?'

'Because,' I venture, 'the person who came out to the lighthouse that night was wearing a diving suit, and the police asked me specifically if I knew whether Bjørkang and Arnt had any reason to take diving equipment with them when they set out that evening. And last, but not least: no one apart from the two of them knew that I had found the body. The only logical conclusion would of course be that they had something to do with her death, and didn't want anyone to know about her. Isn't it?'

'So why have they gone now, then?' Harvey probes.

'Yes, that's what doesn't add up in all this,' I complain. 'Why the hell haven't they come back yet? The problem is that if they don't turn up soon, then I'm in trouble. A whole shitload of trouble and grief, do you get it?'

'Because?'

'I have a feeling that the police are cooking up a scenario in which the policemen came to the lighthouse that night when I was there, and that they subsequently disappeared. I'm the last person on earth who ought to be anywhere near a case involving two vanished police officers.' I give a heavy sigh as my stomach pains resume, and double up in agony.

'Are you OK?

'My stomach,' I gasp through clenched teeth. I have already poured three full measures of Duphalac down my throat on the trip back from Tromsø, without effecting any change down below.

'Tough shit.'

'Quite literally, yes.'

'Talk to Merethe. I know she often helps a few of the residents when they're suffering from constipation.'

'Is she at work now?' I ask, squirming as the pain sears through my gut.

'Yes, she's there till nine tonight. You'll find her in the canteen or at a therapy session,' Harvey says as I reach the apartment door.

'By the way,' he adds as I fumble for the keys. 'The daughter of the couple who lived in the apartment before you is travelling north for the funeral on Wednesday. She intends to clear it out and take their belongings back south with her again. If you could—'

'Of course.' I squat down as another bout of pain grips me. 'Speak later.' I round off the conversation and head towards Skjellviktun Residential Centre's main entrance. It's time to find Merethe.

CHAPTER 32

There is complete silence in the corridor. No wheelchair or walking-frame users, only the sound my shoes make as I pace along the newly washed linoleum floor.

The canteen is crowded. With extra chairs added, lots of residents are milling round the tables, but no one is speaking and the room seems charged with an anxious stillness.

At one end of the room, several tables are pushed together. Two people sit there facing the gathering. One is the nurse called Siv, holding a microphone in her hand. The other is an old man with a dusty brown toupee, in a short-sleeved Hawaiian shirt and wearing a gold watch that glitters in the light from the fluorescent tubes directly above.

'Thorkild,' I hear a voice whisper, and I catch sight of Johannes at the same table as before, sitting beside Bernadotte and Oline in the wheelchair.

'And you are?' Bernadotte hisses as I squeeze into a seat just as Oline lays a cold hand over mine: 'My goodness, how you've grown. I remember when you were little and Agnes used to take you down to the shop to buy ice cream. Good heavens, how you yelled and screeched if you didn't get what you wanted.'

'Here you are.' Johannes pushes one of the four bingo cards he has across to me, together with a red felt-tip pen.

'Two rows,' he says, nodding at the cards before I have time to protest.

'The number is seven. Seven. Seven.' The crackling loud-speakers are on tables flanking both walls.

'What did she say?' Bernadotte whispers. 'Did she say thirty-two?'

'No,' Oline answers, slightly agitated. 'Seven. We already had thirty-two a minute ago.'

We hear the metal cage holding all the balls rattle again before another ball rolls out.

'The number is forty-three. Four three. Forty-three.'

'Have you seen Merethe?' I ask, as Johannes uses the pen to mark the card. He is engrossed in the game and sits with his glasses perched on his nose, staring down at his three remaining cards.

'We'll have a break after we've gone through all three rows,' he replies swiftly without looking up from his card.

'The number is twenty-one. Two one. Twenty-one.'

'Bingo!' Oline screams, grabbing hold of my arm.

'What?' I stare at her as if she is having some kind of attack: she is clawing at my arm as she waves her free hand above her head. 'Bingo, bingo!' she yells before starting to haul me up by the arm. 'Look, look, move yourself, lad. You've got bingo!'

Putting her glasses on her nose, Siv peers out into the cafeteria as Oline drags me out of my seat. 'Was that bingo?'

'Yes,' I answer in a hoarse voice, flapping the card as if it were a dead fish. 'Apparently so.'

I get to my feet and wait for the man with the gold watch who is working his way towards us. 'Fine,' Siv intones into the microphone when he finally reaches us: 'You can start.'

The man snatches the card and lets his index finger follow the numbers until he arrives at the place where two of the rows have been completed: 'Five, twenty-one, thirty-two, sixty-six, eighty-two.'

'That's all good.' Siv's voice rasps over the row of loudspeakers. 'Then the next one.'

'Seven, forty-three, fifty-seven, seventy-two, ninety.'

'That's OK too,' the loudspeaker announces. 'Then we can get ready for three rows. Full house.'

The man with the gold watch hands me the card and returns to the prize table, where he takes his time examining each individual item before finally selecting what he has been looking for and returns to our table.

'Here.' He hands me a pack of fifty Christmas napkins decorated with elves in a variety of sledging exploits.

'Thanks a million.' I give a deep bow and sink into my chair as soon as the handshake is over. All of a sudden I feel Oline's hand on mine. Her head is shaking slowly from side to side. 'Clever boy,' she says, before crouching over her own cards once more.

'Then we're off again,' Siv announces over the loudspeakers, followed by more rumbling of balls and several numbers read out: 'Seventy-seven. Seven seven. Seventy-seven.'

I shove the packet of Christmas napkins over to Johannes, who shakes his head and sends them back to me. 'You keep whatever you win.'

'The number is fifty-nine. Five nine. Fifty-nine.'

'Bingo!' Oline shouts for joy and grabs hold of my arm again. 'He's got bingo again!'

'Hell and damnation,' an old man exclaims, thumping his hand on the table and flinging down his pen.

I glance at the card and see that she is right. The third row is also complete. Reluctantly, I stand up again, looking lost, as I scan the assembled residents, while the bingo inspector sets off on his expedition across the room yet again. Siv gives me a nod of acknowledgement, and fiddles with the microphone as if it were something altogether different.

The procedure of reading out the numbers is repeated before the man takes off again and starts to rummage around on the prize table, eventually returning with a cake tin that he sets down reverentially on the table in front of me.

'The truffle cake?' Johannes says, with a grimace. 'I thought it was the cream cake for three rows now?'

'Don't you start, Johannes,' the man with the watch remarks. 'That comes after the break, you know that.'

Oline leans over to me: 'Yes, Agnes will be pleased, you'll see.'

'Now we'll take a short coffee break before we make a start on the final game of the evening,' Siv announces into the microphone. She stand up and crosses to a long table where members of staff are setting out big silver pots of coffee, bowls of sugar cubes and jugs of cream.

'I think I'll leave now,' I say, pushing the card and cake over to Johannes. Since Merethe is not here, I decide to go to the apartment for another dose of Duphalac followed by some severe straining on the toilet. 'Thanks for the hospitality.'

'The cake's yours.' Johannes passes it back to me again. 'I don't like chocolate anyway.'

'What about you?' I look despairingly at Bernadotte and Oline, who both promptly shake their heads.

Oline pats me on the shoulder and whispers: 'Take care, my boy. And say hello to Agnes from me.'

I pick up the pack of napkins and the cake tin and head for the exit. Behind me the sound system crackles and something that sounds like a gramophone record wheezing out organ music – from an ancient harmonium that must belong to the era of prayer meetings – comes on and a girl with a southern accent bursts into song ... '*Aunt said the angels asked them to come home ...*'

'Escaping?' a female voice enquires once I have managed to close the door behind me at last.

'I've been looking for you,' I answer as I catch sight of Merethe approaching with a net bag filled with small coloured stones in one hand.

'Well, here I am,' she says, grasping my arm with her free hand. 'What did you want?'

Chapter 33

'Have you been like this for long?' We are walking towards the residential centre exit, Merethe still holding me by the arm, as if I am one of the residents here and have to be led back to my room for a nap.

'Just over a week,' I answer. 'Since I got out of prison.'

'Prison?' I notice her grip loosen on my arm. 'But aren't you—'

'Long story. You can ask Harvey about it some day.'

'It doesn't matter.' She squeezes my arm again. 'I like you.'

'Thanks,' I mumble, glancing down at the bag of stones. 'What do you use them for?'

'These are my crystals,' Merethe replies. 'For the therapy sessions. Harvey buys them for me in Russia.' She squeezes my arm harder. 'Cheaper there.'

'Do you sell them?'

'Of course. Jasper, quartz, green aventurine – we call them wish stones – rose-red rhodochrosite crystals. Deep-blue lapis lazuli from Afghanistan, amethysts, rhodonites, fluorites of every colour: white, black, purple, blue, green and yellow, and yes, some are even almost colourless. Emeralds, malachites, onyxes, carnelians, rubies, agates, apatites, beryls, topazes, sugilites, rose quartzes and moonstones.'

'Expensive?'

Merethe blinks both eyes at the same time. 'At a profit, of course.'

'And these are what we are going to use to … to—'

'Exactly,' Merethe concludes once we reach the apartment.

'Cake?' I offer when we enter the living room. I deposit the container and the pack of napkins on the kitchen worktop.

'No thanks,' Merethe says. 'I heard it was you who won the truffle cake. Berit baked it – she's one of our residents in the open rehabilitation unit.' Merethe opens the bag of semi-precious stones and dips her hand inside. 'But she forgets things. Her cakes are not what they once were, to put it kindly.'

'Brilliant,' I answer, leaning across to the kitchen work-top as a fresh bout of pain starts to spread inside me.

'Are you ready?' Merethe holds two stones from the bag in her hand, rubbing them together. She beckons to me. 'Don't be scared. Crystal therapy is totally harmless. Just bring the blanket and lie down on the floor here in front of me.'

The room is warm and I remove my sweater so that I'm wearing only a T-shirt. I prostrate myself before her, lying on my back.

'Are you taking any medicines?'

I nod.

Merethe rubs the stones together in her hand as she looks at me. 'I can see that.'

'Oh?'

'Miosis. Small pupils.' She nods with her mouth closed. 'It's absolutely fine. It can sometimes help you. Anything else you need to make yourself comfortable?'

'The radio,' I answer in a thick voice, stretching out on the floor while Merethe goes to switch on the radio. 'And the coffee machine. Turn it on.'

'Like that?' she asks once she has put on the coffee machine and found a radio station playing light, dancing piano music.

I nod my head. 'No, wait,' I exclaim, standing up without warning, grabbing the bag from the table and dashing to the bathroom. In there, I grab the perfume bottle from the bag and close my eyes and mouth before spraying a fine mist directly into my face. Then I swallow two more OxyNorm tablets and return to the living room.

Merethe sits waiting beside the coffee table. She wrinkles her nose at the strong smell of perfume, but makes no comment and simply kneels down beside me once I have lain down on the floor again. 'The human body has seven chakras,' she says, heating a red stone in her hand. 'These are energy wheels, or centres, distributed along your spine. These chakras control, radiate and regulate the body's use of energies and impulses. We place the stones on these points so that the vibrations from the stones home in on the same vibration that exists within the chakra.'

Merethe starts to arrange stones of different colours and shapes in a straight line from my middle all the way up to my forehead, where she places an oval, purple stone. Then she picks up an elongated, transparent crystal and moves it in a strange pattern over my stomach and groin area, speaking continuously in an undertone, so quietly that I fail to catch what she is saying.

I feel my thoughts being wrapped in thick, smothering clouds as a feeling of calm and wellbeing falls over me, suppressing the pain and substituting new, improved sensory cells, mystical receptors that only medicines can arouse, and that wish me nothing but good. At last, the drugs are beginning to take effect.

'I felt a great sorrow.' I hear Merethe talking through a white, grainy membrane that has materialised between us. 'The first time we met and you gave me your hand, in my house. That was why I pulled it away so quickly.'

'What?' I say, slurring. My eyelids are heavy and it is hard to keep them open.

I can feel the stones on my stomach, chest, and throat, all the way up to the one on my forehead. It is as if they are spinning, round and round, at the same time gradually eating through into me, bit by bit. Pinning me down on the floor so that I can't move.

'Later, I saw her between you and Harvey, outside in the car park.'

I try to whisper something but my lips won't move.

'She seemed to be floating between you.'

'Frei,' I moan. 'At last.'

'I feel her energy again now,' Merethe continues, drawing out her words, as if they are painful to articulate. 'Stronger, more insistent, as if she's pressing forward and wants to break through.'

'Do you see her?'

'Yes.'

'Tell her I need to speak to her. Tell her I've made up my mind. I'm coming, if that's what she wants.' All of a sudden, I can hear Merethe grit her teeth, so hard that they are grinding together.

'She's freezing.' Merethe's teeth are chattering, and now I feel the room grow colder, as if someone has opened a door and let in the winter chill. 'The darkness, she wants to show me something in the darkness,' Merethe says, shivering, and turns to face the window. 'Out there.'

'Where? At the lighthouse?'

Merethe does not answer: just sits, trembling, on the floor beside me.

'Ask her about me,' I plead. 'Ask her why she won't come any more.'

'She's floating.'

'Yes,' I say dreamily, blinking. 'We're both floating.'

'So dark.'

'Do you see the moon? Do you see it?'

'It's cracking,' Merethe whispers. 'I hear rumbling and beating on metal all around me. My God, I'm freezing.'

'The crash,' I gasp. 'Oh my God, the crash!'

'Water. I can hear it all around me; feel the taste of salt well up inside me.' Unexpectedly, Merethe starts to cough, long, heaving bursts, as if she is being choked to death. 'So cold,' she says, shaking, as she gasps for breath.

'Let her in. You have to let her in so that I can talk to her.'

'Oh my God!' Merethe suddenly touches her chest as though struggling to breathe. Her voice has changed, and there is something in it, something that wasn't there a few seconds ago. 'This isn't how it's supposed to be,' she groans, her eyes darting around the room as if searching for something.

'No, wait,' she adds, still trembling, 'I don't understand what you're saying.'

'Frei?' I cry, 'Frei, are you there?'

'No, no.' Merethe whimpers at my side. 'I don't understand.' I feel her hand fleetingly brush my arm. It is as cold as ice.

'Frei?' I try to get up, but my body is nailed to the floor, and I can't move a muscle. The sense of calm within me has dissipated, and panic is starting to spread through my body. 'Why don't you answer?'

'Hush.' At once I am aware of Merethe's frozen hand on my arm again. 'You have to stay still, Thorkild.'

'What? No, let me go. Frei! Wait for me, I'm ready. Ready to come with you.'

Merethe's ice-cold hand slides over my arm and shoulder before she presses the flat of her hand on my lips. 'Now I understand,' she whispers, closing her fingers on my mouth. 'It isn't Frei. It's somebody else ...'

Chapter 34

Merethe's hand has drifted from my lips and is trailing awkwardly on the floor as she kneels in front of me. In the end I manage to wrench myself out of the medicine-induced stupor in order to sit upright.

Merethe gazes at me with a faraway expression, then gets to her feet and turns to the curtains behind the sofa, where she whispers into the darkness: 'Who are you?' She reaches her hands out as if she were blind and has just encountered a stranger. 'Why are you here?'

Cold air seeps from her lips and nose. She rocks gently from side to side, shaking her head and squeezing her eyes shut. 'I don't understand.' Her eyes open again and more cold breath billows from her mouth.

'What does she say?'

Merethe shakes her head once more. 'I don't know. I don't understand.'

I move to stand up, but Merethe holds out a hand to keep me seated.

'She is pointing at you,' Merethe tells me. 'I can't understand what she's saying. She's talking gibberish – or some different language.'

'Frei.' I stare at the curtains behind the sofa where Merethe's eyes are pinned. 'Is it you?' I stagger to my feet and skirt round the table to the window. 'Please say something. I can't wait any longer. I need an answer. Do you want me to come?'

A bitter taste of salt water spreads through my mouth and down my gullet as I approach. The cold draught hits me full in the face as I make a move to catch the curtains and pull them aside. Simultaneously, I hear Merethe break into screams behind me.

I wheel round to see her chest heaving noisily as she gasps for breath. Her eyes are black with fear, her body shaking.

'What is happening?' I ask in desperation. 'I don't understand what's happening!'

Now she is tensing again, thrusting out her chest and spreading her arms. '*Mne xólodno.*'

'What?' I put my hands on her shoulders. 'What did you say?'

'*Mne xólodno!*' Merethe repeats, her voice rising. '*Mne xólodno! Mne xólodno! Mne xólodno!*' I try to place one hand over her mouth, but she slaps it away and starts to scratch and scrape the inside of her mouth with her fingers as if there is something in her throat that she is desperately trying to remove. Her face is drenched with sweat, seeping from every pore.

'Loser!' Merethe spits out. Fear fills her eyes, and a rasping, gurgling noise rises in her throat. She wheezes and gasps for air, as if drowning in her own saliva.

'Stop!' I beg her, struggling to pull her fingers from her mouth. '*Mne xólodno! Mne xólodno!*'

All at once I see her. Not Frei, but a woman without a face or jawbone, merely a hood of flesh surrounding a bare skull stained green with algae. She twists and turns, weaving in and out of the black shadows in Merethe's terror-stricken eyes as she screams at the top of her voice.

'I see her!' I exclaim in amazement, and stumble back. 'My God, I see her! There, inside your eyes. She's here. The woman without a face. She's here!'

The screaming stills. Her hands slump down to her sides. Merethe seems to be about to say something, but instead she simply stands there, her body swaying gently from side to side.

The next moment she collapses in a heap, but her head strikes the edge of the table as she falls.

I scramble across and, as I go to pull her on to my lap, I see there's something wrong with her jaw. I try to close her mouth, but her lower jaw simply falls back into this freakish, gaping position. I spot a gash in the skin beside her ear, bone gleaming white through the tear.

'Wait a minute, Merethe,' I whisper, ready to run for assistance, when a loud knock sounds at the door.

It swings open to reveal Siv, the nurse, standing there with a younger care assistant. 'What has happened? The residents in the apartment next door phoned to complain about a disturbance.'

Merethe whimpers in pain when I release her jaw to sit up straight. 'Phone for a doctor!' I cry. 'Now, at once!'

'What have you done?' Siv asks in horror, as the care assistant takes out a mobile phone.

'She fell and dislocated her jaw,' I answer. Siv dashes across to help me turn Merethe into a stable position, on her side, on the floor.

'They're coming right away,' the care assistant tells us as he returns his mobile to his pocket. He kneels down beside Merethe and Siv, while I haul myself up on to the sofa, fish out my mobile and call Harvey.

CHAPTER 35

'Harvey, you've got to come!' I say, frantic, when Harvey picks up the phone. 'Something's happened to Merethe.'

'Merethe?' I can hear the noise from the boat engine being turned down before his voice comes back again, louder, sharper now: 'What do you mean?'

'She's had a fall, here, in the apartment. We've called for an ambulance. They're on their way. Hurry.'

'I've just been out to the mussel farm. Don't leave before I get there.' The engine roars again and Harvey hangs up.

I set the phone down on the table and sit watching Siv and the care assistant on either side of Merethe. Siv has fetched a blanket and tucked it round her, as well as folding a towel snugly under her head so that her jaw is resting on the soft padding. Her entire lower jaw seems to have been torn out of alignment, the bone unhinged from its socket and now held only by facial tissue.

Merethe's breathing is a husky gurgling, as though her breath is constricted. Sometimes she whimpers softly. Siv is holding her hand and the care assistant gently strokes her hair. Ten minutes later, we hear a vehicle turn in outside and brake to a halt just in front of the apartment. Soon Harvey rushes into the room, with dark, scared eyes, and throws himself on to his knees beside Merethe.

'How did it happen?' he asks, sitting with his back to me, leaning in close to his wife.

'We had a crystal-therapy session,' I start to explain. 'Merethe was going to help me with my stomach problem, as you suggested, when she, we, suddenly saw something, a person, who ... who—'

'A séance?' Harvey abruptly turns to face me. 'You held a séance?'

'Y ... yes, I think so,' I answer in confusion. 'I'm sorry, Harvey. I didn't know that ... She fell. Just suddenly collapsed and hit her head on the table.'

Harvey sits staring at me for a long time, until in the end he forces the shape of a smile. 'Thorkild, take it easy.' He turns to his wife again and kisses her forehead, before whispering: 'It'll be fine. Do you hear me? We'll get through this, honey.'

In the distance, sirens wail. Soon the car park is filled with powerful beams of light. The care assistant sprints outside to welcome the paramedics and returns in an instant with a man and a woman carrying a case and a stretcher between them.

'Her jaw is dislocated,' Siv tells them. 'You can see the mandibular condyle sticking out here.' She points at a wound above Merethe's ear and the man nods before placing his hand warily on the patient's shoulder. 'Hello, are you awake?' he asks, squeezing her shoulder gently.

Merethe moans at his touch and then falls silent again.

'OK then,' the paramedic says, withdrawing his hand before turning to face Siv again. 'How long ago?'

'About twenty minutes.'

'Fine. And she's been unconscious the whole time?'

Siv glances briefly at me before nodding her head.

'OK.' The male paramedic begins to examine Merethe's head and neck, while the woman checks her chest, stomach, upper arms, thighs and legs. At each point they stop and ask whether she feels any pain, and Merethe whimpers softly in response before they continue.

The moment they're done, they carefully place a compress on the wound at her ear and the man shines a light into her mouth to examine the bleeding. Finally, he checks her pupils before sitting back and looking in our direction.

'We'll give her pain relief before moving her on to the stretcher. It's important to keep her lying in a stable position on her side, to check the bleeding in her mouth and make sure that nothing enters her throat. We don't want to intubate because of her jaw, but we have to be prepared to do that if anything should change as we move her.'

Once the woman has administered the painkilling injection, the man turns to Siv and asks: 'Are you a nurse?'

Siv nods in reply.

'Could you hold her jaw while we move her?'

Siv nods again.

'OK.' The female paramedic takes out two smooth boards that she carefully places between the floor and Merethe's body so that they can shift her in a single move straight across to the stretcher from where she is lying.

A fine string of blood hangs from her mouth as they lift her across to the stretcher. On the towel and the floor where Merethe has been lying, I can see a larger dark patch of more coagulated blood. The paramedics arrange for Siv to accompany them in the ambulance to Blekøyhamn and on by helicopter to the University Hospital in Tromsø, and for the care assistant to pass on information to the staff of the centre about what has taken place.

'I'll phone the police from the ambulance,' Siv says, her stern look alternating between Harvey and me. 'I don't know what this guy' – she points a stubby finger at me on the sofa – 'has been up to here, but he's obviously high on something or other, and ... and—'

'What on earth do you think I've done?' I ask.

'Look at her!' Siv yells. The paramedics also stop in their tracks for a second and they all look straight at me. 'Do you mean to tell us that she just fell?'

'Relax,' Harvey says, interposing between Siv and me. 'I was the one who asked him to visit Merethe. He's got stomach problems, she was going to help him, that's all.'

Siv is about to say something, but instead takes a deep breath and starts over again: 'Considering everything that's happened here lately, I'd prefer to phone them all the same. Anyway, I'd like to bring this up again, Harvey – we can't rent out care-home accommodation to just anybody and whenever it suits us. I've made that clear to you before now. Get him out and away from here. As soon as possible.'

Siv marches the two or three steps across to the stretcher, where she and the paramedics hoist it up carefully to the proper height and set off. As soon as they have left the apartment, Harvey approaches me.

'I ... I ...' I start to speak, and stand up.

'My God, you saw her,' Harvey says, gasping. He pulls me into his arms, grabbing my T-shirt by the neck, and gives me a big hug. 'Did you see how scared she was?'

I lack the energy to say anything, and simply hang apathetically in Harvey's arms as I struggle to shut out what has just occurred. Finally Harvey lets go and stands in front of me, swaying. 'They just phoned me.' His voice is hoarse, and only a whisper.

'Who?'

'Police Headquarters in Tromsø. They want me to come in tomorrow morning. At first they asked if I could come this evening, but I managed to postpone it till tomorrow.'

'I see,' I say, suddenly exhausted, as I wipe my face with my arm.

'They want to know about you, Thorkild. About when you turned up here, where you were living, and how you seemed when I took you out to the lighthouse.'

'So they're buying the scenario with Bjørkang and Arnt coming to the lighthouse the night I was there,' I clarify. 'And that something happened to them.'

'But for God's sake,' Harvey mutters.

'Yes, it's crazy, isn't it?'

'Well,' he says. Outside, we can hear gravel crunch as they trundle Merethe's stretcher across the car park to the ambulance. 'I hadn't intended to say anything, but now I feel I must.'

'What is it?'

'Not long ago, when you phoned – I was on my way out to the mussel farm, and had only gone a kilometre or so beyond the lighthouse.' He hesitates. 'Well, when I passed the island on the way back, I spotted something.'

'What?'

'A figure. Or maybe two, I'm not sure. It was so dark, and I was some distance away. But she was sitting—'

'She?' I note that I have to make a strenuous effort to prevent myself from shaking.

'I don't know who it was, but it was a woman, I'm certain of that. I didn't see her face, just her hair blowing in the wind. I only saw her for a split second before the boat's angle to the boathouse made her disappear. I'm not even sure that what I saw was a woman, and not just flotsam washed up against a wall, but the door to the boathouse was open, and I had this tingling feeling, when I looked inside the opening, that there was someone standing there. Someone hiding just inside, waiting for me to go past.'

We hear the rear door of the ambulance slam shut, and immediately afterwards the lights that had glared through the living-room windows vanish and are replaced by the deep blue of evening. 'I have to go,' Harvey says as he watches the vehicle lights fade. It seems as if he has suddenly been brought back to life and reminded of the perilous state of things. 'I have to collect my son, then we must go to Merethe.'

Harvey leaves without a backward glance. He strides across the car park and clambers into the pickup. As soon as the truck has gone, I open the curtains wide and let the evening darkness into the room. I stand there thinking about how rapidly things will tend to go to hell once you've set out on the wrong track, whether you are sitting at the kitchen table with someone you love, or in your life as a whole, when you're left simply hurtling towards your own destruction but unable to apply the brakes or change course.

The snowflakes whirl around in the light cast by the streetlamp in the car park. Somewhere far out there at the lighthouse, a flock of birds is circling beneath a rift in the clouds, and a slender moonbeam streaks down through the gap to strike the island. All of a sudden, I glimpse a light at one of the windows in the main building, a square lighthouse lantern enclosed by all the cold blackness.

'Harvey is right.' I press my face against the glass. 'She's waiting for me out there. It's time to get away from here.'

CHAPTER 36

I am breathless by the time I reach the car park. I drive towards the boathouses at the foot of the bay and draw to a halt where the local police chief and his sergeant had stood when I first came here to take a look at Rasmus's boat in Harvey's boathouse. As soon as I've parked, I jump out of the vehicle and dash below a wind-warped fish frame, zigzagging between turf embankments and slick pebbles.

The birds in the sky look like marionettes in a shadow theatre as they tilt and dive, slowly gliding on near-motionless wings above the former keeper's residence. I can also see birds in the sea around the lighthouse, black with white neck plumage, bobbing on the waves around the rocks on the shore.

The boathouse is divided in two, with a net room at the far end, where you can see various bits and pieces of fishing equipment, green glass floats and tubs and boxes. There is a green plastic dinghy suspended front and rear from two hooks on the wall. It is so small and light that I manage to lift it off and drag it down to the water without any difficulty. I push off and hop on board.

My back is aching when I finally reach the broken quay. I hurl my bag ashore before sliding from the dinghy and hauling myself up on to the rocks. I jog to the boathouse and open the door: there is no one here, and I carry on across the island. At the foot of the stone steps I pause again to peer up at the lighthouse. The entire island and surrounding sea are bathed in a clear shimmering radiance and the cold silver moonbeams make the whole building shine. I take a deep breath and run on.

Sporadic raindrops and the raw, salt tang of the sea lash me, but I don't bother to zip up my jacket or even tuck my T-shirt down into the waistband of my trousers. The front door of the main building is draped with red-and-white police crime-scene tape. I rip this off and drop it on the ground. As soon as I'm inside, I rush straight for the bar.

'Fuck, fuck, fuck!' I rage as I come to a standstill at the bar counter. A faint oniony odour lingers in the room, an odour that was not present last time, but that I recognise from somewhere else. All the windows but one are covered in black plastic bags. In the centre of the floor, a work lamp sits, switched on, and pointing straight at the uncovered window.

I cross the room to open the window. It is pitch dark outside: the break in the clouds has closed over completely now. 'Too late,' I gasp.

I linger there, looking up at the cloud cover for several minutes while I wait, hoping that the gap will open up again, but nothing happens, and even thicker clouds conceal the earlier ones – there is no moon, no light, no Frei: only an endless black barrier that nothing can penetrate. In the end I close the window and leave the bar.

I know it is time to make another attempt to expel the compacted lump in my gut, and I head downstairs to the disco where, according to the sign, a toilet is located.

The door at the bottom is padlocked and I don't have the key. I turn and hurry back upstairs to the foyer and on up to the floor above. The air here is heavy and muggy, mixed with a foul whiff of singed timber and damp fabric.

I head towards one of the slightly open doors. This room is on the floor of the building that seems to have been most badly damaged by the fire. The roof and walls are demolished, both in the main room and inside the connecting bathroom that contains no furnishings or WC. A number of tools are strewn across the floor in addition to a work lamp and several rolls of new insulation material. This room is the only one on this storey in which Rasmus has embarked on restoration work. The rest of the place seems to have been left untouched since the eighties.

The next room is crammed with furniture: a few old camp beds stacked one on top of the other, pushed up against the outside wall, as well as some bunk beds, all made of bleached pine. The interior walls are stained with soot and the carpet is sodden, with oval patches revealing where the material has burned right down to the chipboard underneath.

I close the door and return to the staircase and corridor on the opposite side. The first room I reach has a sign on the door indicating that this is Meeting Room 1. Inside, the curtains have fallen on the floor and the windows are littered with dead insects caught in countless spiders' webs. The chairs are arranged in horseshoe formation around a table with an old computer.

Inside the next door I find a gymnasium full of exercise bikes, benches, apparatus and weights. Dusty mirrors line the walls. One of the air vents is missing, leaving only an open hole in the outer wall. On the floor not far away I spot a bird's skeleton. I step inside, making my way to a door near one of the gable walls.

A tiny toilet with a single washbasin opens out in front of me. I yank down my trousers and sit on the toilet seat, the cold stabbing into my skin when I make contact. All the same I remain seated and push the door to, just enough to leave a crack of light between it and the door frame as I try to relax my muscles and let gravity do its work.

Nothing happens. Eventually I get fed up sitting here, straining away while the wind whistles through crevices and the freezing cold bites at my arse. 'Fuck's sake,' I pull up my trousers, jerk the door open, and head out of the room in the direction of the stairs leading down to the ground floor. It's time to tackle what I came here to do.

As soon as I'm back in the bar, I cross to the glass shelves behind the counter and pluck a bottle of Smirnoff from one of the shelves. I take out the blister packs and a few loose pills, all shapes and sizes of multicoloured tablets and capsules. 'I hope I have enough,' I whisper to myself, swallowing them down with a bitter, burning gulp from the bottle.

While I stand there, I catch sight of a rubbish bag by the wall and an empty cardboard box that had escaped my attention before. I go to open it and find it contains several pairs of used latex gloves and a few empty glass phials, which, to judge from the smell, must have held Luminol.

Luminol is a chemical that exhibits chemiluminescence. You mix it with water and spray it over a particular area to check for bloodstains with the help of UV light, which explained the plastic bags covering the windows. The iron in blood catalyses a blue luminescent effect that highlights even the tiniest traces of blood, no matter what attempts have been made to scour them away with chemicals.

I search for a number on my mobile and press the green button.

'Gunnar Ore,' the voice at the other end announces in a voice half-curious, half-irritated.

'It's me.'

'Thorkild? What the fu … It's only seven hours since I told you in no uncertain terms to stay—'

'Why have you done a crime-scene investigation out here at the lighthouse?'

'What?'

'You've had crime-scene technicians here, I know you have. Why?'

'I work for Delta now – don't you remember that? Or are you so far gone in your godawful haze that you've already forgotten? Crime-scene technology is none of my business.'

'But you know.'

A momentary pause ensues, during which I hear Gunnar Ore's deep intake of breath before he exhales loudly.

He fills his lungs again. 'Police Headquarters in Tromsø have submitted a request for cooperation to the national major crime unit at Kripos, and they were recommended to undertake a crime-scene examination on the island in connection with the case of the missing police officers. That's all there is to it.'

'Why?'

'Surely you can work that out for yourself?'

'So you're involved,' I conclude, mostly to myself, before Gunnar Ore's voice breaks in: 'Come on. Wake up, man. What the hell do you think? Two missing policemen, a civilian who's disappeared, and now you too, running around up there talking about sea monsters, body snatchers and women without faces. For fuck's sake, Thorkild. This is a ticking time bomb. A potential press sensation, a classic cluster-fuck, and here's your name in flashing lights over the door. Who do you think they call when the name Thorkild Aske crops up and fingers start to itch for panic buttons?'

I cradle the mobile phone in my hands for a long time and end up wiping snot and tears away and whispering into the handset: 'Why do you think the two missing policemen came ashore on the island the same night they disappeared?'

'Come on, Thorkild. What are you up to now?'

I keep an eye out in the semi-darkness ahead of me as I force down a mouthful of booze. 'You've found blood, haven't you?'

Gunnar is on the verge of saying something, but suppresses it.

'Where?' I insist.

Another silence.

'Come on, where?' I jump down from the bar and wander around in the gloom. 'Is where you found it significant? Is that why you can't tell me?'

'They got a result with the Luminol in the bar in the main building,' Gunnar finally explains.

'Blood spatter?'

'No, drag marks between some of the floor tiles beside a seating area.'

Blood spatter shows that the person in question has been murdered, or at least had wounds inflicted on the site, while smeared bloodstains merely signify that the person has bled in that location. I cross to the velour sofa and sprawl there. The floor is smooth and cold, as if recently cleaned. 'Do you know whose blood?'

'No.'

'So what makes you think it comes from Bjørkang or his sergeant?'

'We're waiting for the test results.'

'So you're telling me that at present you know, purely through intuition, that the blood comes from Bjørkang or his sergeant and not from Rasmus or the woman without a face? Or someone else entirely, for that matter. Is that the way you work these days? Come off it, Gunnar. You've found something. Something that belongs to one of the policemen, and with blood on it. Haven't you?'

'As I just said …' Gunnar Ore raises his voice the tiniest notch without losing his cool. 'We're waiting for the results—'

'I saw a ghost today,' I interject before he has finished that hellish power-play mantra of his. I am well aware that he is lying, and won't have our final conversation running purely on his terms.

'Really? Seriously, Thorkild. Ghosts – is that where we are now?'

'I saw her in the eyes of another woman. It was her, the woman without a face, the one I found out here at the lighthouse; the one no one seems keen to take any interest in.'

'Well, as I said' – his tone is harsher now, more determined – 'no one at police headquarters believes this woman you're talking about even exists. People are just pissed off that you're haring around hurling accusations and insinuations at two well-respected officers who have vanished into thin air. That's an all-time low, even for you.'

'And you,' I say in an undertone. 'What do *you* think?'

'Me? Well, I can easily tell you that, Thorkild. I think you're a piece of damaged machinery that should get the hell out of here before you fall down again and can't be put back together this time.'

'Humpty Dumpty sat on a wall. Humpty Dumpty had a great fall.'

'Yes, spot on,' Gunnar answers, though there's no sign that he finds the analogy the least bit amusing.

'All the king's horses—'

'Jesus, Thorkild! Pull yourself together.'

'So what's your theory? What scenario are you cooking up?'

'You know I can't—'

'OK, what do you think my role is in all of this? Can't you at least tell me that?'

Gunnar Ore laughs croakily. 'Take it easy, Thorkild. You're not the type, as I also said to Sverdrup the very first time we spoke on the phone. Despite what happened to you and that girl out on Flyplassvegen. You're someone who runs away, someone who shirks issues and gives up in critical situations. I told him that the faster they get you out of this case, the better for all involved.' He hesitates. 'You're someone who always chooses the easy way out when things go against you, and this time you were on the wrong island at the wrong time, that's all.'

'What the fuck do you mean by that?' The hard stone in my gut suddenly reminds me of its presence, and I grit my teeth as I wait for the pang to subside.

'You know what I mean.'

'My modest bid at early parole via the back door?'

'If you like.'

'I haven't told you what happened down there in the showers,' I goad him further. 'Would you like to hear?'

'No,' Gunnar Ore replies crossly. 'Save it for someone who gives a shit. Just get the hell away from ...' He stops. 'You said *here*,' he adds abruptly.

'What?'

'You said *here*, a moment ago when you were talking about the crime-scene investigations. You aren't at the light-house right now, are you?'

This time it is my turn to refrain from answering.

'Bloody hell, Thorkild.' I can hear a storm brewing at the other end of the phone, but it doesn't matter. At last I feel the effects of the insect eggs and alcohol wash over me, high waves cresting, followed by deep troughs.

'Time will soon be up,' I whisper, tucking the bottle of Smirnoff into the crook of my arm before I find a place to sit on the floor beside the boxes of Murano glass lamps and close my eyes.

'Didn't I say quite specifically that under no circum-stances should—'

'Sorry, boss,' I babble to the best of my ability. 'As you say yourself: I'm someone who runs away, and now I have to fly. Caw! Caw!'

Chapter 37

I'm not sure how long I have been sitting here following my conversation with Gunnar Ore. All at once it feels as if my senses have returned and it dawns on me that I can hear music – grinding, piercing, glassy notes composed of drum machines and 1980s synthesisers, as if I've woken to find a party in full swing around me.

I place the bottle of vodka on the floor, where it topples over and rolls across the boards until it settles. My neck is stiff and painful when I lift my head and open my eyes. The bar is dark and cold, far more so than when I first arrived.

Once I have struggled to my feet, I follow the music out into the corridor, where the volume increases. I reach the landing above the stairs that lead down to the locked disco in the basement when a powerful female voice launches into an accompaniment to the invasive synthpop music, some kind of pleading, yearning ballad of heartbreak.

I head down to the foot of the staircase, where the bomb-shelter door that had been locked earlier now stands ajar. Inside, I can make out the outlines of yellow, green and blue lights gliding across one of the walls. Several glass display cases are hung in a symmetrical row with white signs above, though I can't see what these say or what the showcases contain.

The woman goes on singing, begging her lover to take her into his arms, for this night, for always. I push my way past the door to enter a changing room with an open double door at the opposite end. The changing room comprises empty coat hooks on the left-hand side and two doors marked for ladies and gents. The air is oppressive and stale, and everything looks undisturbed since the yuppies stopped dancing down here almost thirty years ago, way back in the eighties.

I cross to the first glass display case fixed at head height into the concrete wall, just as the drum machine works up to the refrain, the singer begging her lover to tell her she is the only one, to tell her this isn't a game, it's love, that it's for forever.

I can see a bird's nest inside the glass case and the sign above the exhibit reads 'Common Guillemot (*Uria aalge*)'. Two eggs are displayed on a bare stone slab. The eggs are dark green with dark, irregular spots and speckles on the shells. Beside this is a showcase that contains two grey-white auk eggs, with black-brown spatters, laid out on a grooved rock.

I follow the wall past other birds' nests in the direction of the disco. The chequerboard pattern from the changing room continues in here. The rest of the interior is made from steel and glass with pastel colours on the walls and ceiling.

On the ceiling a disco ball rotates, spinning and twirling in time to the music, which has now changed to an old classic I vaguely recall from my teenage years. On the floor, either side of the DJ booth, two smoke machines are periodically spewing thick clouds of smoke on to the deserted dance floor as Rick Astley takes over from the female voice, promising and promising, never to give up, never to run around.

The acoustics, dust and smoke from the machine lend the place a sulphur-yellow hue, with pulsing billows of dust drifting between the booths and the dance floor. From the DJ booth itself, high-frequency flashes of light are strobing across the dance floor. 'What sort of place is this?' I growl, raising my hand in front of me, gazing at its movements – juddering, mechanical gestures. It's as though I'm roaming around in one of my own pill-induced dreams, where everything is floating.

Suddenly I stop mid-movement, just as a fresh puff of smoke ascends from the dance floor. The particles of dust curl and twist around me like a display of Northern Lights. And I have glimpsed something farther inside the room.

Something that should not be there.

CHAPTER 38

The pills and alcohol make my body feel ethereal and detached as I stagger across the floor towards the seated woman who leans against the wall in the farthest booth, close to an emergency exit.

But I know that more is required of me, if I am to accomplish what lies ahead. No rusty water pipes or other defective props are going to spoil the grand finale of this performance.

Two tea lights are burning in jam jars on the table in front of her. One hand is held up at her face, as if she is asleep or has simply turned away from the music, smoke and flashing lights.

The woman with no face is still wearing the flimsy night-dress with T-shirt on top, the same as last time. Her body and head are grey and sheathed in a fine layer of ash that makes her look like a mummy of the kind you see on TV documentaries from Pompeii and other such places.

The candles in the jam jars are guttering, about to wink out, by the time I reach the table. Grains of dust rise from the glass, sparkling and swirling before my face like fireflies before they burn out and disappear.

'Are you waiting for someone?'

She makes no reply and I let myself flop on the sofa, edging my way to the opposite side of the table. The wall feels clammy and dank, as if we are sitting in a house at the bottom of the ocean, with icy waters pressing in, slapping against the exterior concrete walls.

I lean towards her, gingerly caressing the thick, bright-green hair that covers her forehead. It is cold and stiff and crackles the way frozen clothes do when you touch them. A smooth, transparent film blankets her body as I brush away the dust. She is as icy and solid as a slab of fish or meat recently removed from the freezer.

'What are you doing in here?' I lean heavily across the table again, carefully scraping away the dust from the deep-frozen corpse. It is like sweeping soot off an old snowman – the dirt slides off, leaving behind grey-black threads on the shiny layer of ice where my hand has been. Underneath, I see blue-purple skin speckled with bronze post-mortem lividity.

I run my hands over the table between us, scraping decades of dust and dirt on to the floor to reveal the glass beneath. It feels as though I've landed in a place reserved for the dead and dying, a transit station where both the faceless woman and I are waiting to journey onwards, from one state to the next.

All of a sudden the music stops: the sound system has had enough and an uncomfortable silence spreads through the room, broken only by the hum of the motor that drives the disco ball on the ceiling and the sporadic, rasping – almost gasping – noise of the smoke machine.

'Can something have gone wrong somewhere?' I rub the sleeve of my jacket on the glass surface of the table until it is clean enough. 'A crossed connection between you and her?' I empty my pockets of boxes, blister packs and pill dispensers, pick out the tablets I want and deploy them on the table. 'You see – I was to meet someone else out here.'

I arrange the pills and capsules in rows on the table that separates us until in the end they form a word. I point at the name constructed of four pill and capsule letters and look up at the ice maiden facing me. 'Her.' I scan the room, with a smile on my face. 'They've organised this party for us.'

The room is not as murky now that the music has been switched off. I snatch some of the pills that form the letter F, summoning saliva to swallow them down. I'm about to grab a few more when the music abruptly strikes up again. A jarring crackle rings out from the loudspeakers and drums start up, with Rick Astley back to his promises of loyalty and perseverance.

Once again the dust begins to whirl and gyrate around the room, in time to the roar from the loudspeakers. I take out my mobile phone and key in one final number. It's someone I need to talk to before I am ready. 'Thorkild?' Liz's voice is soft, warm, and as always contains a hint of ambivalent anticipation, wondering which Thorkild she is addressing this time.

'Hi, Liz,' I slur, trying to shield my mouth from the dust.

'Where are you?'

'At a disco,' I grunt, grabbing more pills from the table in front of me.

'Sorry?'

'I'm at a disco,' I say quietly, letting my eyes drift towards the motorised disco ball that creaks as it twirls round and round. The smoke machine looks as if it has run out of smoke – from time to time it emits a few feeble spluttering sounds followed by grey clouds of dust that shoot across the dance floor to be sucked into the bigger piles of ancient grime that are swirling around. 'Out at the lighthouse.'

'But … what?'

'Somebody has arranged a party for us here.'

'For who?' she asks, sounding tense. 'I don't understand.'

'A party for the dead.' I guzzle the rest of the F and the whole of the R from the glass table. 'And for those soon to follow in their footsteps.' I fill my mouth with more of the insect eggs as my eyes survey the surreal dance hall swathed in dust and dotted with flashing lights.

'Are you angry with me?'

'No, Liz. I'm not angry.' My body has stopped aching, and I have a strange feeling inside, the one that I had when I boarded a plane for the USA. That had been only a few days after Ann-Mari and I had submitted our divorce papers, applying to have our marriage dissolved once and for all.

The mediation process and period of separation had not helped – nothing would help, we'd known that all along. The one thing separation had achieved was to amplify the distance between us. While I sat there looking out of the window, watching the plane climb skyward before eventually arriving on the other side of the ocean, it struck me that the old Thorkild would never be able to accompany me out of the aircraft when we landed at Miami International Airport eleven and a half hours later. Just as it had been a new Thorkild who had been carried from the showers down in the basement of Stavanger Prison one and a half years later, and yet another who had walked out through the prison gates barely a week ago.

In fact there have been many Thorkilds. Too many. Even before that, from the time when that other person left his first home in Iceland and travelled with his mother and Liz to Norway, to Oslo, such a very long time ago. But this is the final version. The era of all the Thorkilds is over.

'Have I told you about the pipe, Liz?' The dust swirling thick in the room, from floor to ceiling, is stinging my eyes and nose. Somewhere in my diaphragm, the boulder is still pressing with no intention of budging.

'No,' she whispers. 'You haven't explained anything about what happened to you in prison.'

I tilt my head to one side so that I am sitting in a semi-upright position with my face turned to the wall and the woman opposite me at the table. 'Well, then, it's about time,' I say, slurring my words, as I shut my eyes. 'Before the doors close and the band goes home.'

CHAPTER 39

Stavanger Prison, 13 February 2012

Robert, Arne Villmyr's lover and Frei's dance partner, whom I'd met at the dance class at the Sølvberg Arts
Centre, had sent me a letter in which he asked for permis-
sion to meet me after I had gone to jail. I could see him
sitting out in the car park together with Arne the day he
arrived. They'd kissed and embraced for a long time before
he stepped out.

Stavanger Prison had four visitors' rooms. Nineteen of us
were shut up in the section called NORTH, the name they
had allocated to the communal unit. Three of the visitors'
rooms were in use, since it was the winter holiday period.
The only one vacant, normally used for family visits, had
children's drawings on the walls and was furnished with
IKEA boxes filled with toys as well as a small drawing table
for the youngsters. Two big Winnie-the-Pooh teddy bears
were sprawled topsy-turvy at either end of a leather sofa,
holding their arms out as if expecting a hug.

'Are you ready?' the liaison officer asked as we stood in
the doorway of the visitors' room and glanced inside. At
the opposite end of the corridor I saw Robert approach,
accompanied by a prison officer. He held his hands clasped
in front of him and his head was lowered, as if he were a
new inmate on his way to his first encounter with his fellow

prisoners. 'You've got half an hour,' he added. 'Afterwards, a mother is coming with her little boy to meet his father for the first time. We've reserved the family room for them for the rest of today's visiting time.'

'That's fine,' I answered, glancing over at the children's corner, equipped with a blue and yellow mat on the floor under the table with drawing materials laid out and boxes of toys. On the mat there was a picture of two giraffes munching apples from an abnormally high apple tree.

'Wait here,' the liaison officer said, closing the door once Robert and the other officer were so close that I could hear their footsteps just outside.

I crossed to the sofa and sat down at one end beside one of the teddy bears. The next minute, the officer opened the door again to admit Robert. 'OK, boys. Half an hour. I'll be waiting outside.'

Robert paused for a second in the middle of the room as the door closed behind us. He was just as well dressed as the last time we had met, in suit trousers, a white shirt and a knotted scarf under his parka jacket. However, his facial features were somewhat coarser than I recalled, now that he was standing so close to me.

'Just stay in your seat,' he said as I made a move to stand. He untied his scarf and took it off, putting it down on the table together with his jacket. He ran his hands through his thick, black hair and sat at the opposite end of the sofa. He too found it hard to deposit the Winnie-the-Pooh bear down on the cold floor, with the result that he ended up holding it on his knee.

'We haven't really been introduced properly,' he said, with no sign of initiating any kind of handshake.

'No,' I answered. 'We met that time at the dance class, that's all. What do you want?'

Robert nodded silently without responding to my question. In his letter, he had merely written that he had something to tell me, something we needed to discuss. After the accident I had returned to Bergen, to the bedsit where I remained until the court case went ahead. I was suspended from my post in Internal Affairs as soon as the results of the blood tests had come through. The only person I had heard from had been my boss, Gunnar Ore, who phoned to ask me to stay away from the office and my colleagues, and not to speak to the press, before concluding by telling me to go and die in a ditch.

'How's it going with' – he looked down his nose before continuing – 'your face?'

Instinctively, I touched the red half-moon-shaped scar beside my eye. My finger slid down my cheek, where the skin had been torn in a star-shaped pattern, pulling the tear ducts below the surface.

'Everything heals,' I answered, withdrawing my finger as soon as I was aware of it tracing the contours of my face.

'Everything heals, yes.' He shifted position. 'We buried Frei at home in Tananger.' He sat at an angle at the end of the sofa, his legs dangling just above the floor. 'You can go there when you get out, if you like. Arne and the family say that's OK.'

'Thanks.' I reached across and pinched Winnie-the-Pooh's ear hard with my fingers.

'Her parents have left again. Arne is sitting outside in the car – he didn't want to talk to you.'

I nodded without releasing the teddy bear's ear.

'It was me who gave Frei the GHB.' Robert pressed his hands into the teddy bear's belly as it sat on his lap. 'That was why I had to speak to you. To tell you that.'

'Why?' I asked in a frosty tone.

Robert looked at me with a strange expression on his face. 'Didn't you know? Really?' He shook his head. 'No,' he continued. 'I think you do know.'

I shrugged without letting go of the teddy bear. 'Does it matter?'

'No, maybe not.' Robert went on gazing at me for a while before he ploughed on. 'That was why I came. Because I had to tell you that.'

'Have you told that to your boyfriend out there in the car and her parents too, for that matter?'

'They know,' he answered, again flashing that peculiar smile of his, or not a smile really, but something between a smile and a grimace. 'You knew she had a boyfriend,' Robert continued. 'A ... police officer at Police Headquarters here in Stavanger?'

Once again I shrugged my shoulders. Deep inside, I felt my body turn cold. 'Simon Bergeland,' I said eventually.

'Yes, exactly. The officer you were here to investigate in that case of yours. A thief with a propensity for violence, an all-round scumbag whose world was about to come tumbling down around his ears because of your arrival in the city.'

I went on kneading the teddy bear's ear. I could not bring myself to say anything, and just sat there while my insides writhed. 'Why?' I asked.

'Why what, Thorkild?' Robert's half-smile had transformed into a slit devoid of anything other than contempt and loathing, not towards me, but towards us both, that we were the two left sitting here, left behind, while Frei lay cold in a coffin somewhere out in Tananger. He pressed his chin on the teddy bear's head as he bit his bottom lip.

'Why did she do that?'

'At first she was probably simply curious about who you were. I also think that she wanted to know about Simon, what it was he was actually up to. The day after you had met each other at Café Sting, Frei came home and asked if I could get her some GHB.'

'What was the plan?'

'Not really a plan,' Robert said. 'I don't know all of it – I didn't want to know – but you and Frei were to drive into a police road check, high on GHB, and Simon would at least gain some breathing space. Maybe the case would even be dropped, who knows?'

'A police road check.' Suddenly it was there again: that bitter taste of cider in my mouth. 'Really?' I went on, struggling to swallow. 'How original.'

'Simon had sent her a text message saying that there would be a road check on Tanangervegen that evening, and asked her to make sure you drove into that road check, under the influence. She told me that when I arrived with the bottle of cider. My God,' he groaned and put his face in his hands. 'You've no idea how much I regret getting at all—'

'Tanangervegen?' I stopped for a moment before jerking upright. 'What do you mean?'

'The police road check,' Robert answered, looking at me again. The half-smile was back. 'It was at the end of Tanangervegen.'

'No,' I whispered, shaking my head. 'No, no, no!' I glanced up at Robert as I made an effort to restrain what had started to unfold within me. 'You're mistaken. It can't have been on Tanangervegen. It—'

'No, Thorkild.' Robert sat, unruffled, looking at me between the ears of the stout teddy bear. 'I'm not mistaken.' Putting the teddy down on the floor, he approached me. 'But you didn't go to Tananger,' he said, placing his hand gently on my arm. 'Did you?'

I did not answer. My insides felt transparent and fragile. Everything had tensed, and the smallest movement was painful. This was a new type of ache – one I had not known even existed. An ache that does not go away, one that is never going to depart.

Robert was about to say something further when a knock sounded on the door. One of the liaison officers popped his head inside and said: 'Nearly time up, boys.'

I went on sitting without moving so much as a muscle. Robert stood up and put on his jacket and scarf. 'I think that was everything.' He nodded to himself as he pulled up his jacket zip. 'Don't you agree?' He turned to the door where the liaison officer had appeared, before his eyes turned back to me. 'Goodbye, Thorkild. I don't think the two of us need meet again.'

Then he left.

After the liaison officer had escorted me back to my cell, I picked up my toilet bag and towel and headed off to the gym hall. 'How could you?' I whispered to myself as I went through the hall past two men who were dismantling the volleyball net. 'How could you do that to me?' I stopped in front of one of the lockers and took out a skipping rope before moving on to the changing room and the communal showers at the opposite end.

The changing room was empty. A cloud of steam floated over the opening between the showers and changing room, and you could see puddles of water on the floor that separated the benches.

Water trickled down the stone walls and a curtain of vapour drifted up beneath the ceiling where pipes ran from the shower walls across to the other side of the room. I crossed to the door, opened it a crack and peered out. The two men in the gym hall were gone and the main light was switched off. I closed the door and dragged one of the benches with me into the shower.

As soon as the bench was in place, I turned on all the showers, setting them to the hottest setting possible. I jumped up on the bench and lobbed the rope over a thick pipe bolted to the ceiling. I looped it over twice more, then tied a noose at the other end.

I leapt down from the bench and headed across to the showers on my right, which all switched off automatically after one minute, and turned them on again. The older showers on the other side, lacking this automatic function, were still spewing out hot water that splashed on the stone floor as the veils of steam grew denser.

I scurried back to the bench, grabbed the noose with both hands and pulled it down over my head. The noise of the water pounding the floor, bombarding the rusty drains on either side of me, was deafening. The waves of steam clung to my skin, keeping the cold at bay. Condensation closed my pores and insulated the acute, amorphous pain inside me.

I used both hands to catch hold of the rope above the noose, pressing my toes against the end of the bench below me to kick it away. *There you are, Thorkild*, I thought, as my head was pushed forward and down on to my chest and the rope tightened around my neck. *The worst is already over.*

My torso twisted slowly in a circular motion while my legs kicked convulsively out into nothingness. The showers on the right were no longer spilling out water. I remember feeling annoyed that they couldn't keep me company all the way. Suddenly, out of the corner of my eye, I caught sight of the wall clock in the changing room through the mist. A round metal dial with thick hands. Three minutes past five.

Why didn't we go to Tananger that evening, Frei? Oh my God, why couldn't we just have gone to Tananger ...?

CHAPTER 40

'She used you.' Liz's voice jerks me out of the fog of drowsiness I've talked myself into. I open my eyes at the very moment the motor on the disco ball above the dance floor finally gives up the ghost. It's as if all of its innards have churned themselves into thousands of tiny plastic and metal fragments that are now simply drumming around inside the casing. 'She manipulated you and wanted you to lose your job. I hate her,' Liz sobs bitterly. 'Even though she's dead, even though we've never met, I still hate her.'

'You don't understand,' I blubber, struggling to haul myself into an upright position again. 'But it's not your fault. We're both like that. That's why we choose people like Arvid and Ann-Mari, damaged, distant people who can't ever see, who won't ever come close. But that's not how it should be, Liz. I knew about Frei and Simon Bergeland the whole time. Her name was in his case papers. Frei was playing a game with me, and I let her do that. Manipulation and information, don't you see? That's what we do, you know, people like me. Manipulate thoughts.'

'I don't understand—'

'Hasn't it dawned on you yet that your brother is an illusionist? A fucking first-class expert in smoke and mirrors, so utterly engrossed in the role that ... that ...' Having lost my thread in all my gabbling, I wolf down the last of the pills on the table between me and the woman without a face. I only just manage to keep my hands steady as I run them across the table. Several of the pills and capsules clatter on to the floor or drop between my jacket and shirt as I lift them to my mouth.

'Did you ever meet him?' I can hear Liz struggling to grasp the conversation and steer it in a different direction. 'Her boyfriend?'

'No. He never turned up for interview that day. Later I heard that Internal Affairs searched for him for a while, but they never found him. Puff,' I say, laughing. 'Vanished into thin air.'

I give a brief nod of farewell to the frozen woman without a face opposite me before getting to my feet and staggering back across the dance floor. The music has stopped again. The room is totally silent, disturbed only occasionally by the asthmatic breathing of the smoke machine and the whirring of the broken disco ball on the ceiling.

I regain my balance by grabbing hold of one of the pillars that divide the dance floor from the rest of the room, and stand there for a second while I struggle to remain on my feet and strain to see through the clouds of smoke. Ultimately I glimpse the green man on the white emergency exit sign and take a deep breath before preparing to lurch towards the exit.

I try to slow down by using my free hand to take hold of the metal handle, but miss and bump my head on the door without feeling anything at all.

'You mustn't do it,' I hear Liz whisper as I pick up my mobile and stand with my back against the door. 'I can't manage on my own, Thorkild.'

'Don't be afraid.' I cough and choke until I've collected enough saliva in my mouth to swallow. 'The cold metal will melt me down and then, hey presto …' I rub my eyes in an attempt to remove the film obscuring my sight. Instead I end up letting in more soot and dirt so that tears begin to flow, though I don't feel any pain in my cheek. 'Then it'll all be over. No more Thorkilds.'

I can hear Liz breathing through the receiver, heavy and irregular between the words she believes I will hear and the tears she is unable to hold back. However, this is not about us, about Liz and me. It applies only to myself, and the woman who stands waiting on the other side of the rivers.

I wheel round and lean all of my weight against the door handle. At last it gives way and opens with a creak.

Chill sea air pummels my face and tears at my lungs when I inhale. Above me I can see that a break in the weather has begun. I leave through the emergency exit, heading for the rocky shore, and stop in front of a massive boulder. My eyes scan the seascape ahead of me. 'But I didn't see,' I smack my lips in an effort to taste the salt water on my tongue, 'or hear what it was you were trying to say …'

I produce a gasping, gurgling attempt at laughter, but the sound sticks in my throat and I gag instead, over and over again, as I pick my way across the slippery stones between the emergency exit in the basement of the main building and the rocky shoreline. In front of me the waves surge forward with head-shaped tangles of seaweed sploshing and beating against the bedrock.

'So beautiful,' I sing as I teeter down the slick stones towards the water's edge. My face is turned up to the sky, where I can see deep ravines in the incandescence of the moon. Rivers of silver wend their way out of these gashes, imbuing the circumference of the sky with various metallic tones, and an even fierier pattern snakes down in spiral-shaped galaxies closer to the centre.

'Thorkild,' Liz bleats. I can hardly hear her now. The roar of the sea swallows her voice. 'Don't hang up. Please. I promise I'll be a better sister – I won't nag and bother you any more – I …'

I've taken the mobile from my ear as I halt on the edge of the rocks. 'We didn't go to Tananger that evening,' I mumble to the black waters in front of me. Farther out, the rivers merge into the murk, illuminating it from the inside, an intense silver-black bioluminescence. Soon I too will be reshaped and become part of this enormous energy. 'We took another road, didn't we, Frei? A completely different road.'

I lift my arms out at my sides and place one foot slightly in front of the other, retching again. Vomit runs down my chin and clothes, but it no longer matters, since the water will wash it away. This time the water will wash everything away.

Above me, the sky is completely open at last, and cold colours join the two points together. The wormhole is ready. Shutting my eyes, I press my lips together and force my body forward.

I meet the darkness before me with a subdued splash and roll over on to my back where I lie rocking in time to the beat of the ocean, back and forth, as I am drawn away from the rocks and into the silver rain. 'This is the last,' I gasp, as the salt water washes over my body and face. 'The very last Thorkild.'

Chapter 41

My final day with Frei, Stavanger, 26 October 2011

Frei was leaning against the door outside Stavanger Police Headquarters when I descended from the office on the first floor that had been allocated to me. The duty officer had phoned to say I had a visitor. At first I thought it was Simon Bergeland who had finally turned up for interview, and asked the duty officer to let him in.

'What are you doing here?' I asked when I emerged. I rolled my shirtsleeves up to my elbows as I approached her. A gentle breeze rippled over the hairs on my forearm and rustled through the trees outside.

Frei was wearing a short skirt with a black and white pattern, black trainers and a white, loose-fitting, silky sweater with long sleeves. Her eyes were heavily made up and she had something black, resembling a dog's chain, around her neck. 'Shall we go for a drive?' Her lips had thin streaks of deeper red where they met. I was aware of the smell of alcohol on her breath as she moved in towards me and gave me a brief, clumsy hug.

'I can't,' I answered when she withdrew again. Her eyes were red, as if she had been crying. 'I'm waiting for somebody.'

Frei tried to smile, all the time struggling to stay on her feet. 'He's not coming,' she said, laughing, and leaned against me so heavily that I felt her breasts press against my chest.

'How do you know that?'

'I know everything,' she replied, still laughing, and slipped away from me again. She retreated a couple of paces and used one of the lampposts for support as she cradled a bottle of cider.

'I don't doubt it.' I stood gazing at her, aware of the fragrance of her shampoo and perfume clinging to my clothes. I just wanted to succumb to the illusion again, to hell with the consequences, and go over, lift her up and bury my face in her hair.

'Are you coming?'

'Stop it, Frei.'

'Please.'

I looked at my watch. It would soon be six o'clock. The interview with Simon Bergeland was supposed to have started three hours earlier. I knew he would not appear today either, unless he was the person we were travelling to meet.

'Where are we going?'

'Tananger.' Frei tossed the bottle of cider to me.

I opened the bottle and took a swig. The cider tasted sour, almost like battery acid.

'What's going on in Tananger?' I asked, passing the bottle back to her.

'Wait and see.'

I glanced at my watch again before finally answering: 'OK, then. But I'll drive.'

I rang the doorbell and the duty officer let me in again. I raced up to the office, cleared my belongings and packed away the

AV equipment I had fixed up. I tried to call Simon Bergeland's number one last time, but his phone was still switched off.

Frei was standing beside the car, smoking. Getting in without any exchange of words, I reversed out of the car park and drove to Lagårdsveien.

'Am I beautiful?' she asked abruptly without meeting my eye. Her head was turned towards the side window and the lights of the city beyond.

'Extremely. What's in Tananger?'

'You know nothing about me,' she went on without heeding the question.

'Who are you, then?' Following Frei's directions, I turned on to the first roundabout that took us over to the main road and on out of the city on route E39. A strange prickling sensation had erupted in my chest, and was spreading up through my shoulders and face.

'I could have been yours,' she said without warning as she turned to face me. Her eyes were dark, almost as if she was scared. 'Really.'

'No, I don't think so,' I replied, attempting a smile, but it was as though something was tugging at the muscles beneath my skin, holding it back.

'Why not?' Frei took the bottle of cider from the centre console and nervously began to rotate it in her hands. Opening it, she took a mouthful, replaced the cap and continued to spin it between her hands. 'You just said you thought I was beautiful.'

'Extremely beautiful,' I added.

'Well?'

'This is just a game, isn't it?' The streetlights had already come on, and the rain clouds were dispersing above us.

Frei stopped twirling the bottle and let her gaze drift back out through the side window. 'What if there were just the two of us?' she whispered into the glass.

'Are there more of us?' I turned theatrically towards the empty rear seat. 'Hmm, nobody here, you know. Do you see anyone on your side?'

I wanted to tell her that I knew all about her and the corrupt policeman whose career she was trying to salvage, to tell her that I was going to nail that bastard to the wall, spreadeagled, no matter how damn much she had thought to degrade herself in order to save him. But I didn't. There was still something that restrained me. A ridiculous, naïve belief that as long as neither of us paused the game, then everything would stay true. That we hadn't simply been thrown together by circumstances, shadows drifting across each other on a busy shopping street, but that we were two people, a man and a woman who felt a kinship.

'Yes,' I said in the end. 'If there were just the two of us.'

'Even though you know I'm not who you thought I was, that I'm constantly lying, that I use—'

'Yes,' I broke in. 'Every time, if you absolutely have to know.'

I could see her smile through the side window as she continued to roll the bottle of cider between her hands. 'Thanks.'

'Thanks?' All of a sudden I felt very hot – my chest, my neck and my face. Gentle, teasing warmth that tickled its way to my stomach and made me want to laugh. Endlessly. 'What do you mean by *thanks*?'

'I just wanted to know.'

'But aren't there only the two of us, then?'

Frei straightened up and finally looked straight at me. She pointed at my chest. 'One,' she said before directing the finger at herself, 'two.'

'Ah, very funny,' I answered, laughing, as my belly filled with more soft, tingling heat. My hands tensed on the steering wheel. The cool leather increased the tingling in my hands, making it more acute, as if something beneath my skin was trying to force its way through and out.

'Wow.' She sat watching me with her mouth half-open.

'What?' My grip on the steering wheel became harder. It felt as if my body would glide up from the car seat, that we were two astronauts on a training flight on board some sort of NASA spaceship in parabolic orbit.

'Did we just spend a moment together?'

'Did we?'

'I think so,' she said, laughing.

'By the way, tell me when we need to turn off. I—'

'You've already driven past,' Frei said, 'a long time ago.'

'Fuck,' I muttered, trying to look for an exit road or another roundabout, all the while struggling to concentrate on sitting still in my seat. 'Should I turn round?'

'No.'

'Weren't we going to Tananger?'

'Don't turn.' Frei rested her head on my shoulder. 'I want to drive on.'

Above me I caught sight of a dark hole in the steadily contracting rainclouds. A sharp light had appeared up there. Not the steel-blue light from the sky that you are used to at this time of year, but something warmer, more earthly, as if I was staring at a glowing silver disc. Eventually, as the clouds went on receding, I could see that it was the moon.

I was about to tell Frei that it appeared to be bleeding, that thick streams of silvery lava were flowing out of its glittering surface, but before I managed to say anything there was a hollow thud from somewhere underneath me. The next minute I was flying, hovering over the driver's seat with my hands clutching the steering wheel. Frei also took off, and I could see her terrified face turn away from mine, and her hair rise to form a halo around her head. Right after that, a loud bang sounded and I could feel my head thrown forward to smash into the wheel.

The last thing I remembered was the bitter, metallic taste of cider in my mouth. It was still there when I woke the next day and they told me that Frei was dead.

TUESDAY

CHAPTER 42

This certainly isn't a new Thorkild. That is the first thing to strike me when I wake. Dead stars glint like cats' eyes from the night sky above me. All around I hear the water splashing and lapping as I pitch up and down to the rhythm of the waves.

I am lying on some sort of raft made of seaweed, carrier bags, tail-ends of rope, plastic containers and other rubbish drifting around in the surf out at sea. I turn my head and can see that this raft is the wreckage of the quay torn loose from the lighthouse by the storm on my first day there.

'My God,' I groan before glimpsing a young bird a couple of metres away – it too seems to have settled on this island of trash as it drifts through the night. The fledgling is plump, speckled white and dark brown, with a black beak. It resembles an eagle as it sits there staring at me through its large, piercing, raptor's eyes before turning away to draw its wings more snugly around its body and press its head down on its chest.

'Help,' I whisper, twisting my head towards the bird. 'Can you help me?' The next moment the pains stab through me – first my neck and cheek and then my belly: the throbbing internal pressure in my digestive system, heavy as an anchor, is back with a vengeance.

I turn my gaze to the sky and squeeze my eyes shut as convulsive pains shoot through my gut. When I open them

247

again, I can see that one of the cats' eyes has tumbled from the sky and is on its way to earth, like a falling star.

I break into howls just as the young sea eagle struggles to flap its wings.

Gradually, as the falling star approaches, filling the heavens with light, the bird gives up and wraps its wings around its body, lifts its head and opens its beak, either in a final attempt to defend itself or because it thinks it is about to be fed.

I make a move to raise my hands and shield myself from the strong light, but don't have the strength. One hand is caught in something. I turn my head aside to discover what it is and catch sight of a grey hand protruding from the forest of seaweed and bladderwrack beside me. The white, stiff fingers are closed around mine, as if in a welcoming handshake.

I touch the hand sticking out of the tangle of weeds with my fingertips. The surface feels cold and smooth, like a tailor's dummy. I pull my hand back and try to free the other one from its clutches when all at once I glimpse the falling star again.

It is poised above the surface of the sea like a fire-breathing dragon, sending foaming surf in all directions, and making the whole island of rubbish rock heavily from side to side. I grab the bottom layer with my free hand and yelp in fright and pain as scraps of plastic, foam and ice-cold ocean spray rain down on us.

Soon there's a splash not far from where I am lying, and I can see a black figure carving its way through the water towards me. The baby eagle has rolled itself into a ball of brown and white feathers with only the beak visible. The rest of its head is buried somewhere inside. 'Let go,' a voice yells at me. A man has appeared by my side.

I shake my head and cling more tightly to the floating debris and the ice-cold human hand.

'Let go,' the man repeats, pointing at a metal stretcher lying in the water beside him. 'I can't get you on board the helicopter if you don't let go.'

I turn my gaze back to the eaglet. I can make out one of its eyes under the feathers, before it hisses at me. Then its head vanishes back under its wing at the same time as the man in black at my side starts to force my fingers open, one by one, from whatever I am clutching under the surface of the sea.

'Let go! Do you hear? You have to let go, or I can't get you up in the basket!'

Eventually I capitulate – I don't have any energy left, and the man begins to drag me by the shoulders, away from the island of waste. At my back, an entire human body emerges from the seaweed I'm towing in my wake through the flotsam towards the metal stretcher.

'You go up first,' he says calmly, and I nod a response. 'Then I'll take him next. OK?'

In the end I let go, and my rescuer rolls me over into the basket before attaching the harness around my body. He waves to the light and remains behind with the dead body. The very next minute, I feel myself being hoisted out of the water and towards the light, while my teeth chatter and my lips quiver.

As they haul the basket into the helicopter, I also catch one final glimpse of the baby eagle down there on the quay debris wrested from the lighthouse a few days earlier. It is on its feet now, flapping its wings as foam and seawater whipped up by the rotor blades cascade over it.

'Hey! Can you hear me?' a member of the helicopter crew asks, as he wraps me in blankets and aluminium foil and the basket is winched down again.

I try my hardest to say something but realise that I can't manage to gather enough air in my lungs. The anaesthetist touches my cheeks with his hands. Only now do I notice how cold I am. I can barely feel the heat of his hands through my own skin. 'Hey, you can't close your eyes now,' he adds as his fingers pinch and knead my face.

The roar of the helicopter surrounds us, drowns us in the pulse of the rotor blades. Soon the winch comes up again. The body in the basket is removed and set down beside me. It is wrapped in blankets and foil by the time my rescuer comes on board, slamming the door behind him.

'Have you seen this?' the crewman unfastens the strap on his helmet and puts it aside.

'What?' the anaesthetist turns to face him. 'Is it one of the missing policemen?'

'No,' my rescuer replies coolly. 'I guarantee you it's not.'

The anaesthetist pulls away the blanket so that the corpse's head beside me falls to one side, and we lie there face to face: the head is swollen, as if the flesh and muscle attachments have loosened from the skull. The face is peppered with starfish, and the eyes no more than two gaping holes with tiny creatures gorging themselves inside: void of thoughts or words or anything else that Anniken Moritzen and Arne Villmyr are so desperately pining for where they sit in Stavanger, still waiting.

'There's more,' the crewman says.

The anaesthetist takes hold of Rasmus's head, turning it away from me again, and looks up at the crewman. 'More? What do you mean?'

'Can't you see it?' the crewman says in a sharper tone of voice. 'It's attached to the body with cable ties.'

'Bloody hell!' The anaesthetist gasps in astonishment.

'I didn't see it at first either.'

'What is it?' The anaesthetist shakes his head in confusion and turns to face me again. 'Hey, you!' Angrily, he clicks his fingers in front of my eyes several times in succession. He moves over so that I can see the arm he is brandishing. I can see the radial bone and forearm bone where they protrude from the grey flesh with thick, rough fingers spread out, looking for all the world like an inflated glove. 'Whose arm is this?' he demands irascibly. 'Do you hear? What kind of bullshit is this?'

I don't have the strength to answer. Instead, the noise of the rotor blades spins a thread that binds me to the cold, the pain, and all the rest. Soon it is totally dark.

CHAPTER 43

Next time I wake, I am lying tied to a bed that is being dragged out of the helicopter and rolled across the tarmac from the helipad, into a doorway and on through a long, narrow corridor. The journey ends in a room with a sterile odour, crammed with apparatus, cables, gadgets and people who all seem to be waiting just for me.

The bed is placed in the middle, between the machines. A guy approaches and leans over me while two other personnel cut my clothes open – one from the neck down and the other starting on my legs.

'Hello! Can you hear me?'

I only just manage to open my mouth, but can't utter a sound.

'Do you know where you are?' The man uses two fingers to pull my eyelids open and shines a light on my pupils.

I shake my head.

'My name is Dr Berg. I'm in charge of the trauma reception facility here at Tromsø University Hospital. If you can manage to say anything, shout out. Try to stay awake. OK?'

I attempt to nod but am struggling to keep my eyes open.

Dr Berg turns to a man nearby and exchanges a few words with him before turning his attention to me again. 'Do you know why you're here?' He lays two fingers on my chest and taps with the other hand before repeating the procedure on the other side. Afterwards he turns to a third person standing in front of a whiteboard directly above the headboard. 'Pulmonary oedema, doubtful.' The man at the whiteboard scribbles something on the surface. 'Pneumothorax, negative, but take an x-ray all the same.' Dr Berg's focus turns to me once more. 'So,' he says in a quiet, deep voice. 'How are you feeling now?'

'Rasmus,' I whisper, just as a young man with short hair hauls my wet, sliced up clothing away from under me before starting to fix electrodes to my body. 'Where is Rasmus?'

'Martin!' Dr Berg barks without looking up from my face. 'Hypothermia?'

'Mild to moderate.'

'Temperature?'

'32.3 degrees Celsius. Rising.'

Dr Berg grasps one of my hands and squeezes the fingertips, one by one. 'Can you feel that?'

I nod and he takes hold of the other hand to repeat the process. 'Does that hurt?'

I nod again.

Letting go my hand, he leans in closer. 'Do you know your name?'

'Thorkild,' I gasp, struggling to rise to get a grip of where I am and what all these people around me are actually doing. 'Thorkild Aske.' The man at the whiteboard writes my name in block capitals at the top of the board.

'Catheter?' asks the nurse at my side, who is now finished with the electrodes and has followed the cables back to a machine that suddenly starts emitting noises.

Dr Berg nods his head before turning to a young blonde woman who has approached the bed. She stands beside the nurse with the electrodes. She has a long, thin rubber tube in her hand that she gives to the nurse before grabbing my penis and indicating that she wants the tube back.

'Just relax,' the nurse with the electrodes says. 'It'll soon be over and done with.'

'What? My life?' I try to say, but the words are drowned in saliva and slime. Around me the trauma team is in full swing, checking the oxygen saturation level in my blood, my pulse, blood pressure and temperature, all the time reporting back to the man who is filling in the details on the whiteboard. The woman with the rubber hose has now pushed it so far up my urethra that only a small tail is protruding from the opening. She pumps air or fluid in before moving off to her next task.

'How did you end up in the sea?' Dr Berg is poised above me again just as a heavy x-ray machine is hoisted over my chest. 'Did you fall in?'

I shake my head. 'The rivers.'

'The river? Did you fall into a river?'

Shaking my head, I try to turn away, but Dr Berg wrenches me back. 'Listen to me. How long have you been in the water?'

Once again I shake my head.

'Do you feel any pain anywhere?'

'My stomach,' I answer. Dr Berg plants himself in front of me and places a hand on my abdomen.

'Here?' He looks up at my face.

I bow my head when he touches the block of stone. 'It's stuck,' I whisper.

'I see.' Dr Berg addresses the man at the whiteboard. 'Constipated.'

I try to close my eyes to shut out the noise, but every time I do so, Dr Berg is there with his fingers, forcing them open again with a fresh round of questions. I observe sensation in my body slowly but surely reviving and the pain receptors being reset.

A lab technician enters with a sheet of paper for Dr Berg once the x-ray machine has been pushed away. They stand there talking in whispers for a few seconds before the technician departs again. Dr Berg moves over to my clothes and goes through the pockets. After a short time he returns to me, laying his hand on my forehead and pulling my eyelids wide open.

'Have you taken any of these?' He holds up a fistful of empty blister packs, containers and dispensers for tablets and capsules. 'Have you?'

I attempt to close my eyes, but Dr Berg pulls them open again as he reads out the names on the boxes and jars to the man with the pen. 'Which ones? Sobril? OxyContin? OxyNorm? How many?'

I succeed in forcing my eyelids closed for a moment before Dr Berg's fingers are there once more. 'How many?'

'All of them,' I answer with a sigh.

He lets go, taking a step back, and beckons to the lab technician at the door.

'OK, guys,' I can hear him say. 'I think we have all the damage vectors on the board now.' I've already closed my eyes and squeeze them so tightly shut that Dr Berg will have to use both hands to force them open. 'This is a suicide attempt. So,' he claps his hands before he goes on, 'let's double-check the x-rays for haemorrhages and possible cardiac tamponade and finish off here. After that you can transfer him to the intensive care unit.'

I feel the people around me pull back, one by one, with the sound of rattling equipment and slicing rubber echoing off the floor. Suddenly Dr Berg's voice booms out again: 'Everyone ready? Martin, reference temperature? Teemu, Janne, you take care of the reports and wheel the patient up. The rest of you, get ready for the next stint. The night has only just begun, people.'

CHAPTER 44

I usually dwell on these thoughts. In them, I am with a woman – I don't know who she is, because I can't see her face. We are in an apartment, and I am me, now, only without the damaged face. We are sitting – no, circling each other in a bright room with high windows. She is wearing one of my shirts and skipping barefoot across the warm pine floor, as though we are in some corny yoghurt commercial. It is so real, this dream sequence, so intense, that when I wake, or drop back into reality, it is as though I have just walked through the wrong door and my body is filled with panic. I turn round and grab the door handle, but there is no longer any door there. Just me, on my own, in some anonymous, sterile waiting room.

Ulf calls it an alter ego. He says it is a fantasy I return to when I want to die, my paradise, and my seventy-two virgins – the alibi I give myself when life gets too complicated. Ulf is right about many things, and almost certainly this too. The problem is merely that he doesn't understand. How difficult it is. He doesn't appreciate that the abyss is not something that you flee to, but something you run away from. Emptiness is not a cold body in a coffin beneath the ground. Emptiness is me, Thorkild Aske, in this apartment with the high windows. Alone.

I wake to find a young, fair-haired nurse standing over me and fiddling with my genitals. 'Take it easy,' he says softly as

I startle and try to pull the quilt back over my crotch. 'I'm going to empty the catheter and take it out. You're being transferred to the renal unit. Then we'll give you an enema and see if we can sort out that stomach problem of yours.'

It is morning, or at least light outside. The square room contains four beds, two on either side. My body feels numb and the absence of pain makes me fearful.

'Give me a mirror,' I say when the nurse has removed the tube from my urinary tract and dropped it into a metal bowl shaped like a sauceboat.

'A mirror?' the nurse peers suspiciously at me as he pulls off his gloves. 'What do you need a mirror for?'

'I need to look at my face.'

'We can't give you a mirror.'

'Why not?'

'Because,' the nurse answers, before seeming to arrive at a compromise that will ensure our time together is as brief as possible. 'I can push your bed closer to the basin on the way out and then we can raise you up in the bed, if you promise not to get up to any mischief.'

'Mischief? Such as what?'

'I think you know what I mean,' he replies, grabbing the side rail control that elevates the top of the bed.

The scar that runs from my eye down my face is almost gone, masked by my grey complexion. Even my lips look anaemic and transparent, as on a dead fish. 'The same,' I whisper, turning away.

'What do you mean?' the nurse asks, adjusting the bed to negotiate the doorway.

'My face,' I say, with a sigh. 'It's the same.'

'What had you expected to see?'

'A new Thorkild,' I murmur, moving on to my side, though the pains in my belly suddenly prevent me from completing the action. I remain on my back, staring at the lights on the ceiling, as the nurse wheels me along corridors and into the lift en route to my next stop.

'You'll be sharing a room,' he tells me, drawing to a halt outside a door. 'At least until this afternoon. Try to relax.'

A bald man in his seventies sits in a bed by the window. His complexion is pale, and his sinewy hands lie like broken twigs in his lap. He is gazing out at something on the other side of the glass.

'How long do I have to wait here?' I ask as the man by the window turns to face us. He lets his eyes slide slowly and apathetically over us, then turns his attention to the outside world again.

'We're keeping you here for observation until your kidney and liver function are back to normal.' He gestures towards the bathroom door. 'Well, if you just go in there and wait, I'll come and help you,' he tells me before crossing to the washbasin.

I sit on the toilet-seat lid and wait. After a while, I hear a tentative knock on the door, and then the nurse is standing in the doorway with a smile on his face, an enema in one hand and a tube that most resembles a fully-grown adder curled around the other.

'Ready?'

'Can hardly wait,' I say, and shudder as he places the enema on the edge of the basin and dons a pair of gloves. He spreads a towel over the floor between the shower and toilet.

'Just lie on your side on the towel with your back to me.'

I kneel on the towel and pull down my boxers before stretching out on the warm floor with my hands beneath my cheek for support.

'By the way, my name is Jens,' he says. I can hear his breathing grow more laboured. 'It might feel a bit uncomfortable to begin with,' he tells me, guiding the end of the tube in. 'But I'll be as gentle as I can.'

'Mhmm,' I gasp, squeezing my eyes more tightly shut and pressing my knees against my midriff. The sense of dying, socially, has never been stronger than at that very instant.

Jens speaks with a remarkable, ludicrous, naturalness. He holds the plastic bottle of enema in one hand and the tube in the other. Sometimes he squeezes the enema bottle slightly to produce more liquid, other times he rests his hand on his head as he talks, telling me about completely trivial things such as the water level in the city at present or the approach of winter, as I answer in words of one syllable. It is almost as if we are two old acquaintances, brought together in this comical teamwork without it changing anything between us.

'There!' Jens lowers the arm with the enema bottle and places the bottle on the floor behind my back. He puts one hand on my naked backside. 'It's important that you hold tight for as long as you can. Preferably for five or six minutes after I remove the tubing, so that the oils have time to work.'

Jens pats me on the buttock before putting the plastic bottle, tube and gloves into a white plastic bag that he twists and ties. He lifts the toilet lid and says: 'Just stay lying there and squeeze as long as you can. When you can't hold out any longer, jump on to the toilet and let the forces of nature do their work. I'll wait outside.'

'Yes, OK,' I comply, gasping and choking back tears, as I struggle to keep the other end closed.

As soon as Jens has shut the door behind him, I can hear the radio being switched on and the volume turned up. A man sings and the music builds up note by note to the impending storm.

At first next to nothing happens, apart from me feeling the oil press against the block of stone at one end and the opening at the other.

The pressure and the impulse to relax my diaphragm muscles intensify, matching the tension inside. I feel like a balloon gradually filling up with air, but have already decided to hold it in until the song is finished. If for no other reason than that I have lost all sense of time in here, lying with all my focus on this squeezing exercise.

At last something happens. A grumbling is starting to grow, first only as a tiny little prickling air bubble, but this is different, it doesn't stop, doesn't move quick as a flash like the others have done, staying there in limbo and causing pain. This one travels downwards and I can both hear and feel it.

Immediately afterwards it is as if something is released, a shifting, like when a glacier several million years old yields to pressure and slides several millimetres closer to the fjord and the meltwater at the base of the valley.

… *Why don't you love me any more?* The vocalist on the radio demands bitterly and heartbreakingly while I slacken the muscles in my legs enough for them to slip down a little, stretching out a notch to feel the pain…. *Tell me, why don't you love me any more? … Tell me why don't you love me any more?*

Once again I feel the glacier move. This time I also have to release some of the squeezing pressure, just enough so that a thin trail of oil trickles out and runs down my skin towards the warm towel. I adjust the squeezing exercise as I stretch out my legs a little more, so that I can turn my head sufficiently to ascertain where the toilet seat is located.

When the refrain concludes, the synthesiser assumes control so that the melody almost twists in on itself. I can no longer manage to hold it in. I quickly roll right over on my back and then on to my stomach before hauling myself up on to the seat and flopping down on the toilet just as the oil shoots out of me and washes down the inner walls of the porcelain bowl.

The next moment, I feel that the glacier has loosened. The ancient mammoth, led down by the forces of gravity, moves slowly but surely through slick, newly oiled paths towards the exit. A crescendo is reached, evacuation is in full swing, and the song on the radio ends with a scream. 'At last,' I whisper, and press my elbows harder against my haunches.

CHAPTER 45

'Did it go OK?' Jens puts his head round the bathroom door and eyes me with a sympathetic expression as he taps his fingers on the doorframe.

'I've survived,' I gasp from my stance in front of the mirror, rinsing my face with cold water.

'What now?' I ask when I re-enter the room where the radio is now switched off.

'What about some rest? You must be exhausted after all you've been through in the last twenty-four hours.'

'And afterwards?'

'Afterwards a doctor will come to talk to you about next steps,' Jens says, smiling, making a move to go.

'My medicine,' I exclaim, anxiety prickling as it dawns on me that he is about to leave without handing over my insect eggs.

'Yes?'

'I need something to help me sleep. I—'

'Let me discuss that with the duty doctor,' Jens says and departs.

I retreat to the bed and sit down with my hands on my forehead like blinkers. After a few minutes, Jens comes back with two 10 mg OxyNorm analgesics.

'Here,' he says. 'Now you've got something to help you sleep.'

I lie motionless in bed once Jens has left. In my dozy, apathetic condition I am unexpectedly aware of a sound, two short peeps followed by a deeper, sharper metallic noise from the floor somewhere just beyond the mothership of my bed.

Immediately afterwards the sharp peep is there again.

I roll over on to my side and try to peer over the edge of the bed, where a tiny green light is blinking inside the plastic bag stuffed full of my sliced-up clothes.

I crawl forward until I reach the bag and can empty out the contents on the floor. The air is suddenly filled with a foul stench: the smell of salt sea and wet fabric pushes into my nose and down through my respiratory system. My mobile phone, miraculously, has outlasted its stay in the sea.

I lean towards the edge of the bed and bring my mobile underneath the quilt. The battery-charging light is flashing, signalling that the phone is about to give out. The first peep I had heard was obviously the alert, because I had received a new text message, and the next had indicated that the phone was about to run out of juice. The message is from Gunnar Ore, and just short enough for me to manage to read it before the mobile switches itself off: 'Formal interview tomorrow at three o'clock. Chief Interviewer from Kripos on his way.'

In other words, the police have received the results of the tests they carried out during the crime-scene investigation at the lighthouse. In that case, these must confirm that the blood they found comes from either Bjørkang or his sergeant. That they have already summoned a chief interviewer also means that they have constructed a satisfactory sequence of events, and agreed that now is the strategic moment for further questioning of the suspect.

I let the phone fall to the floor and turn over on my side with the quilt over my head. 'Now it starts all over again,' I murmur to myself. 'This time they're going to crucify you if you don't do something.'

I am on the verge of closing my eyes when another wave of oxycodone hits the stimulation centre in my brain and pulls me inwards. I throw off the quilt and clamber carefully out of bed and approach the man at the window. The tingling in my hands and toes tells me that my pain receptors are wakening again.

'Hey, you, do you have a phone I can borrow?'

The man gazes mutely at me for a long time until, in the end, he points to a jacket hanging on the footboard of his bed. I approach and find a mobile phone in one of his jacket pockets. 'Thanks, it'll be a short one,' I assure him, as I bring the phone across to my own bed, and sit down to key in Anniken Moritzen's number.

'I found him,' I whisper as soon as she answers. 'I found Rasmus.'

'Yes,' she replies in a hollow voice. It sounds as if she is in her office, despite the news. That is if she doesn't have the same machines grumbling and groaning in the background at home as well. 'A man phoned this morning. Where have you been? I've been trying to call you all day long.'

'Did he give you his name?'

'His name? What do you mean?'

'What was the name of the man who phoned?'

'I'm not sure,' Anniken says, hesitating.

'Could it have been Gunnar Ore?'

'No, I think it was Sverdrup or something like that. Maybe Martin. Why do you ask?'

'OK, forget it. It doesn't matter. What did they tell you?'

'That they had found a man in the sea, and they thought it was my son.'

'What else? Did they say anything else?'

'They asked if Rasmus had been with anyone else while he was there.'

'And had he been?'

'Yes, some of his friends had been there earlier in the autumn, but they left long ago.'

'Did they mention a woman?'

'Yes. They wanted to know if there was a girl among them, or if he had met anybody up there. I told him that he had been alone.' I can hear Anniken let out a sob at the other end. Her voice is muffled, as if she is holding the handset close to her mouth. 'What was it that actually happened?' she finally whispers.

'I'm sorry,' I answer, 'that I couldn't do any more.'

'No, wait, Thorkild,' Anniken sniffles at the other end. I cut her off and click off the phone. There's nothing I can do for her now. She will have to walk the rest of the road herself. That is the only way. The old man at the window glances at me without moving his head.

I get to my feet and walk across to his bed. 'I can't bring the dead back to life again, can I?' I say, returning the phone to his jacket pocket. 'Not me – I have a hard enough job looking after myself.'

I lie looking up at the ceiling as I try to conjure up the man and woman in the apartment with the high windows. Instead, a bird of prey appears, and shortly after that a hand materialises between us. It grabs hold of me and refuses to let go, no matter how loudly I yell and how hard I fight to free myself.

I open my eyes and sit up in bed, gasping for air. The man at his post by the window is still sitting on the edge of his bed, gazing dreamily out through the glass. 'Fuck it,' I growl, and pull off the quilt. 'Fuck, fuck, fuck!' I rip off the tape holding the cannula on my hand, yank out the device and ditch it. Then I head for the door.

I have to know what it was that I found out there.

CHAPTER 46

I open the door and peek out into the corridor. My whole body is shaking uncontrollably, the hairs on my arms are bristling, and I can feel the cold from the floor slide like icy fingers up over my bare skin. All I am wearing is a plain hospital nightshirt that buttons down the front.

To my surprise I find that the mortuary is not situated in the basement of the hospital. On the contrary, it is on level seven, only two floors below where I am now.

The lift stops, the doors slide open with a rattle and rumble, and I walk towards a door marked 'A7 Clinical Pathology'. Inside there is another corridor with a reception and duty room. Farther along I see a number of doors on both sides, and I can hear voices and idle laughter from the duty room. Reception is deserted.

I sneak in and continue on along the corridor until I reach another door marked 'Post-Mortem Room' and 'Mortuary'. Inside I find a changing room with a wardrobe and a shower. On the opposite side, a door leads into the autopsy room itself and the mortuary refrigerator. I can smell the unmistakable odour before I have even opened the door.

I come to a standstill in front of the wardrobe. Beside it I spot a big shelf for shoes with three pairs of boots on it. I open the wardrobe doors and pick out some green scrubs from inside of it then put them on over my nightshirt. I also pull on a pair of the boots from the shelf in an attempt to fend off the cold from the floor outside, and cross to the door to the autopsy room to open it a crack.

The room is bright, with white walls and shiny metal equipment. There are saws, bowls, basins, scales, tongs, scissors and knives. In the centre of the room, a group of ten or twelve people are fanned out with their backs to me round one of the autopsy tables.

I am about to turn away and close the door when I hear a voice behind me: 'Are you late too?'

I peer into the face of a young woman with frizzy blonde hair pulled back in a knot. She is standing in the changing room with a camera slung over one shoulder, buttoning her green coat.

'I'm Astrid,' she says. 'Police crime technician.'

'H ... hi,' I stammer, clutching the doorframe to have something to hold on to.

'Yep, I've got to attend the post-mortem with you lot.'

'Fine,' I answer, struggling to control my shivers.

'You're one of the students, aren't you?' She approaches after pulling on a pair of boots, and stands in front of me. 'My goodness, you don't look too good,' she says, shaking my hand. 'Do you think you can cope with this?'

'Apologies,' I start to explain, 'I'm really not on top form today.'

'Is it the smell?'

I give an exaggerated nod, and Astrid leans towards me with one hand on my chest as she whispers: 'I'm afraid I have to warn you right now. These sea corpses are always the worst.' She winks conspiratorially and pats me lightly on the shoulder. Then she opens the door to the autopsy room and leads me towards the group standing waiting inside between the metal benches.

Astrid grabs two hairnets, face masks, gloves and shoe covers from the baskets just inside the door and hands one set to me. She walks on past the students to stand on the other side of the autopsy table beside the pathologist and the autopsy technician.

I head over to the huddle of students and take a place at the very back. I have observed a few such post-mortems in the past, the first one when I was just a police trainee. The smell and anticipation of what is taking place is the same: your senses do get used to what is in store, and you have to find the proper distance to stand.

This time I regret it at once as soon as I catch sight of the black body bag on the autopsy table in front of us. My legs feel numb and the weight in my gut is aching. I notice how profusely I am sweating under the hairnet and on my temples.

It takes a strenuous effort to stay upright. The sour, bitter smell from the cadaver in the bag fills me with an absurd sensation that I am about to see my own scarred face underneath the zip.

'My God,' I gasp, bumping into one of the students in front of me. The student shrugs and changes position, something I can at least take as a sign that I am still a human being, and not merely an aimless, ethereal, pain-filled substance.

'Are we all present?' The pathologist shakes hands with the crime-scene technician before passing a fractious glance over the assembled company. 'Astrid is from the police. She is here because it is suspected that this may not be a death from natural causes and the police have requested a forensic autopsy of the deceased.' He scratches his nose with the back of his hand before rounding off: 'You'll soon see why.'

'Can I ask a question?' enquires a tall, skinny lad with a Trøndelag accent.

'By all means,' the pathologist answers curtly. 'Ask away. That's why you're here.'

'I've read that there will soon be robots that can conduct virtual autopsies, and that these will soon replace the traditional—'

The pathologist raises his hand, tilting his head slightly to one side, as if he finds the question amusing: 'Look, my dear student.' He pauses briefly then continues: 'Smelling, cutting and holding give us medical insight that no machine will ever be able to replace. However, I'm not someone who stands in the way of progress, so who knows?' He looks across at Astrid, who is now preparing her camera for action. 'Perhaps you have something to add here?'

Astrid looks up from her equipment and shakes her head.

'Well,' the pathologist says, taking hold of the zip on the body bag and starting to tug it down, 'then we're off. And remember, if anyone feels unwell, just take a seat or go outside if you wish. The canteen is an excellent place to re-evaluate your career choice, and decide whether you'd rather have a nice tidy office job.'

The stench hits as soon as he starts to pull on the thick zipper. The plastic creaks, and several of the students sway slightly and start to breathe through their mouths, leaning against one another to keep their balance. The face that comes into view is yellowy and has swollen since the last

271

time I set eyes on it. The skin is thick and doughy, with goose pimples on the cheeks and around the mouth.

'So,' the pathologist says, once he and the autopsy technician have removed the bag to reveal the prostrate body in front of us on the table. Rasmus Moritzen is wearing a diving suit that has burst in a number of places and is bulging in the stomach area and around the thighs. Anniken was right, after all – he had gone out diving when he disappeared.

The students are standing totally still, staring at the same thing as me – the torn-off forearm attached with cable ties to Rasmus's wrist. 'We'll come to that later,' the pathologist comments as the technician uses a showerhead to rinse out the bag. He folds it up and puts it aside on a nearby metal table. 'What are the two most critical questions we have to ask when it comes to diagnosing a body recovered from the sea?' The pathologist looks at us while Astrid takes the initial photographs.

One of the students holds up her hand: 'Finding out whether the deceased was alive or dead when he ended up in the water.'

'In other words, to see whether drowning is the cause of death. Good. And how do we find the answer to that?'

'By examining the circumstances surrounding the death,' the student answers. 'That will tell us when he landed in the water in addition to the discoveries made during the actual autopsy.'

'Clever girl,' the pathologist replies joylessly before turning to face the crime-scene technician, who has by now finished taking preliminary pictures. 'Anything you'd like to add, Astrid? From a crime-scene technician's point of view?'

'We look at the factual and technical evidence, then compare that with the autopsy evidence to see whether they coincide.'

'And what does the investigation say about this case so far?'

Astrid hangs her camera over her shoulder again. Her lips have a slightly peachy colour that glistens in the powerful ceiling lights. 'The deceased disappeared last weekend,' she begins. 'It was assumed that he had succumbed in a drowning accident. Last night the body was found floating in the sea in a clump of seaweed and flotsam, together with another person who is now in the intensive care unit here at the hospital.'

'Alive after such a long time? How is that possible?' the pathologist asks.

'No, the corpse was found by someone who had attempted suicide – he claims he jumped into the sea last night. We don't know yet whether there's any connection between the two.'

I take a step back and out of the group of students as I search for something to sit on. My muscle tremors are harder and harder to control, and I feel dizzy, with an insistent pressure at my temples and in my stomach. I draw a stool towards me and am about to sit down when I hear the pathologist's voice: 'And then we have this.'

I take a deep breath and stand up again, this time leaving the white stool beside my legs so that I can sit down again if need dictates. The pathologist's eyes scan the gathering before stopping at me: 'As you have probably already worked out, the corpse is tightly tied to a foreign object, using cable ties.' Cautiously, he picks up the grey-green forearm lying on top of Rasmus Moritzen's cold body. 'It looks as if it once belonged to a young woman, and after a rapid visual estimate it appears to have spent a long period in the water.'

He waves the arm in such a way that the wrist takes on a feminine curve just as Astrid is about to snap a photograph, and then he puts it back in place beside Rasmus on the metal table. 'A suicide candidate found barely alive, the corpse on the table dead for over a week, and a forearm from an unknown third person who has presumably been in the water for at least a month. Explain that,' the pathologist demands triumphantly before winking at Astrid, who also pulls a smile.

'Do we know who she is?' I ask.

'No,' Astrid replies. 'The police know of no one reported missing who corresponds with ...' She glances down at the arm for a moment and then looks back at me, 'the profile.'

'Was the suicide candidate attached to the two others with the same cable ties?' the lanky boy asks.

'No. They were all found floating in the middle of the fjord on the remains of a quay that the storm had torn loose. If the police had not already had a rescue helicopter in the area searching for the two missing police officers, the attempted suicide would also be lying on the table in front of us.'

'OK, I've got it,' the student says. 'The suicide candidate jumped into the sea, but found the body and changed his mind.'

'There are some aspects that link the candidate to the deceased,' Astrid says. 'But I can't go into that here.'

'I know,' gasps a young ginger man in front of me with red and yellow pimples on his neck. 'The suicide candidate murdered the owner of the arm first, then this guy here, and then dumped them both in the sea.'

'Oh, a murder mystery even before the pathology examination has begun.' The pathologist sucks in his cheeks. 'Well, someone will always want to start at that end.'

I sigh with relief when I finally feel the cold steel of the stool on my fingertips. Letting my body sink down, I bury my face in my hands as Spotty Neck concludes his idiotic analysis to everyone's entertainment: 'He couldn't live with himself after what he had done, so he jumped into the sea after them.'

'Right, that's enough fun for now,' the pathologist breaks in, before addressing Astrid: 'Anyway, what does the suicide candidate say himself? Do you know?'

'The person in question hasn't been interviewed yet,' she answers. I press my fingers into my facial skin and feel the pain receptors in my cheek switch on, one by one, until the heat sets my whole face burning. 'But as you can see,' she goes on, 'he has a bit of explaining to do.'

'OK, students. Enough chatter and chitchat for now.' The pathologist claps his hands quietly. I take one last deep breath and rise from the stool again as the pathologist leans forward, poised above Rasmus Moritzen's dead body. 'Let's get the autopsy under way.'

CHAPTER 47

'We'll start by recording what the deceased is wearing.' The pathologist hovers over Rasmus Moritzen's face. 'In this case it's simple. The deceased is wearing a dry suit. Without flippers or any other paraphernalia.'

The suit has tears and gashes from the feet all the way up to the neck. Astrid, the crime-scene technician, takes pictures and writes key words as the post-mortem progresses, while the pathologist measures and examines. At the back of the head, a cluster of pale, lifeless starfish has gathered around an indentation in the skull. The skin there is greyer, almost brown, and the hair is torn off, or scraped away.

'What is it?' Spotty Neck is pointing at the same indentation in the back of the head.

The pathologist looks up at him. 'A head injury,' he answers sourly, keeping his eyes on the student for several prolonged seconds. 'We'll come to that shortly.' He returns to the injuries on the back. The pathologist confers with his colleague as he works. Once they are satisfied, the body is turned over on to its back again and they set about removing him from the diving suit.

It is like watching a doll with limbs made of sand-filled cloth bags. Arms and legs dangle from the table as the pathologist and autopsy technician struggle with the suit. In one instance, they are forced to place the body in a sitting position. Rasmus's head falls back heavily, his mouth wide open, and his jaws rock gently from side to side as they pull the suit off him and roll it down to his waist. At this point one of the starfish comes loose and slides through his sodden hair to slap down on to the metal bench. The autopsy technician picks it up and disposes of it in a yellow trash container at the base of the autopsy table.

After they have laid the body down again, the pathologist raises the head carefully and places it in the same position as before, slightly upturned and with the mouth open. The diving suit and underwear are carefully removed, leaving Rasmus completely naked. A further round of measurements and photography ensues, from the legs upwards. Every single mark and tear is recorded and catalogued.

When they reach the head, the autopsy technician detaches the last of the starfish around the indentation on the back of the head, using a large pair of tweezers, and drops them one by one into the trash bucket. Underneath them we see a rupture in the skin and the bare skull.

'Ante-mortem injury?' Astrid leans over the wound with her camera and takes three quick photos from different angles.

'Possibly.' The pathologist rinses the wound with the showerhead before using a sponge to wipe it clean of algae and other forms of sea life. 'It's hard to tell them apart in instances such as this, especially when it comes to head injuries. We'll see what the tissue samples tell us.'

'Could it have been caused by a boat-engine propeller?'

'No,' the pathologist answers. 'That would have produced a greater number of larger lacerations in the area of the head and neck.'

'What does the soft tissue injury tell us?'

He presses his finger on the wound. 'Even surface, no fracture on the cranium itself. That indicates a simple, short-lived impact. Any possible coup and contrecoup injuries would tell us more about whether the injury stems from a fall or if the body was stationary at the time of the energy exchange.'

'First impressions?'

'An injury from a blow,' the pathologist murmurs. 'But we'll know more when we have removed the brain. Well, boys and girls – let's press on and open him up. Here we use the en masse method and lift out the organs in the chest first, followed separately by the stomach organs.'

The autopsy technician picks up a sharp scalpel and stands close to Rasmus's torso, stretched out on the table in front of him. He begins by cutting a deep incision along his collarbone, from shoulder to shoulder. Afterwards he takes hold of the skin and pulls it away so that the neck, as far as the chin, is cleared of skin, and streaks of greyish-yellow subcutaneous fat and pale-brown flesh emerge.

The technician continues by making another incision on either side of the body, from the armpit to the iliac crest. The ribs are then cut open with a pair of rib shears, thick curved pliers, before he leaves the shears on the autopsy bench and lifts off the whole of the ribcage. A pungent, pervasive odour seeps out from the open chest and wafts towards us.

I can see how he ties the major arteries before the chest organs are lifted out in one piece: lungs, heart, liver and gall bladder, pancreas, spleen and kidneys end up on the board, where they are separated before the surface, cut surfaces, structures, arteries, lymph glands, fibrous tissue and nerves are examined in minute detail and the organs weighed.

Afterwards samples are taken for cultivation, chemical analysis and other tests.

'OK,' the pathologist announces, lifting the lungs and holding them up with both hands. Bloated, they have a marbled, doughy appearance and are grey-blue in colour, speckled with dark red, and we can see a clear liquid ooze out when the pathologist squeezes them slightly. 'What is drowning?'

'Suffocation with the nose and mouth held under water,' a diligent student in the front row pipes up in a high-pitched voice.

'Water?'

'Any liquid,' counters another.

'Explain the mechanism involved in drowning,' the pathologist continues as he permits some of the students at the very front to run their fingers over the surface of the lungs he is holding in his hands.

Spotty Neck leaps in: 'With the nose and mouth under water, or another liquid, it sets off a fight to surface in order to be able to breathe again. The struggle subsides in the end as the subject tires, and drowning begins as soon as you can no longer manage to hold your breath. Then the liquid will be inhaled, usually accompanied by severe coughing and vomiting, something that will result in losing consciousness followed by death in the course of a few minutes.'

'There is no specific recipe for drowning,' the pathologist comments. 'So that makes it important to search for signs to indicate whether the deceased was alive or not when he entered the water, as well as ruling out any physical factors along the way. Well, then – indicators for drowning. What should we look for?'

'A white froth in the air passages and in the mouth and nostrils can tell us whether the deceased was alive when he sank under the water,' Spotty Neck says proudly.

'Well, my dear student –' the pathologist gestures towards the body '– would you be kind enough to come forward and demonstrate?'

The student takes a step forward and bends towards Rasmus's corpse, which lies there with open chest, empty eye sockets and gaping mouth. It is hard for me to picture the boy in Liverpool shorts standing at the barbecue with his mother and grandparents. *This isn't him – this is something else altogether*, runs through my mind as Spotty Neck puts his hands on his knees and inclines his head towards the cadaver. First he peers down into the mouth before twisting the head and neck so that he can access the nostrils.

'For those of you who actually read the set texts,' the pathologist begins to explain after watching the student in front of him, who has spent some time demonstrating an impressive number of awkward head positions in an attempt to look into the deceased's air passages. 'It is fairly evident that this exercise is a waste of time for everyone involved. Decomposition destroys this froth in the air passages after a short period and replaces it with an acrid red-brown fluid filled with gas bubbles, a sort of pseudo-froth, which our good friend here is inhaling while he makes a fool of himself in front of the rest of the class.'

Spotty Neck swiftly straightens up and retreats shame-facedly into the rank and file again.

'You could check whether the lungs are full of water?' another student suggests.

The pathologist nods. 'But the liquid in the lungs is of the same character as what will result from, for example, pulmonary oedema following cardiac arrest, overdose or head injuries. So what else?'

'Foreign bodies in the air passages, lungs and stomach.'

'They indicate drowning, yes, but water will percolate in through the gullet, windpipe and the major respiratory passages after death also, and in addition smaller amounts will find their way down into the gullet and stomach. But not all the way into the alveoli, perhaps?' He uses his thumbs to squeeze the lungs he is holding, as if feeling his way through the tissue for something deep inside. 'There,' he says eventually, 'feel here.'

The students at the front lean forward and run their fingers over the area the pathologist has marked with his thumbs on each lung. 'And what are the alveoli?'

'Air sacs in the bronchial tubes that allow oxygen to move into the bloodstream.'

'Sometimes you can identify it from the lungs, but you normally have to examine a segment of lung under a microscope in order to confirm the finding. I have my suspicions,' he says slyly, before dropping the lungs into a bowl together with the rest of the chest organs.

The pathologist now turns to Astrid, who is making notes on a writing pad. 'Anything you'd like to add so far?'

Astrid shakes her head and the pathologist nods to himself. 'Well, then, on to the stomach area.'

The autopsy technician makes the necessary incisions before lifting out a quivering lump of jelly comprising the stomach and intestines, which he places on a long workbench beside the corpse. A cloying, sickeningly sweet stench of blood and excrement mixed with the rotten smell of sea and algae invades my senses, making it difficult to breathe even through my mouth.

'This is the most salient part of the autopsy,' the pathologist says. He has produced a scoop and litre measure and hands these to the technician, who at once begins to scoop the reddish-brown liquid from the stomach cavity into the measure while the pathologist approaches the clump of intestines: 'Little water in the stomach can indicate that the deceased died quickly, or that he was already dead when he fell into the water.'

The pathologist takes out a scalpel and separates the stomach from the intestines, closing the severed ends with clips. Then he opens the stomach by making another incision and empties out the contents into a bowl.

The stomach is full of a greenish, bronze-coloured fluid containing a number of tiny molluscs. Still alive, they spin around in the bowl. 'Here the evidence speaks for itself,' he says, going on to examine the organs on the bench, handling them with the same dedicated precision as the ones from the chest cavity.

'How's it going so far?'

'Fine,' the students reply in unison, before the pathologist takes a step to one side and looks directly at me.

'And what about you? Quiet as a mouse in the back row?' He points at me with a bloody finger. 'Are you fine as well?'

I nod my head in response.

'Any questions so far?'

I shake my head.

'Unwell? Tired? Hungover? Or just one of those days?'

'One of those days.'

'Well, in our line of work we can't have days like that. They don't exist. Understood?'

I nod, and the pathologist turns his attention back to the class as a whole. 'First the brain, then lunch.'

He gestures to the technician, who rinses his hands and the yellow washing-up gloves thoroughly under the tap. Then he puts on protective goggles, unhooks an electric saw from the wall and takes up position at the head end of the autopsy table.

'Keep your distance while he is sawing,' the pathologist advises, and the entire group retreats a number of paces while the technician tests the saw. Astrid also takes a step back and uses her hand to shield the camera lens.

The autopsy technician puts down the saw again and makes a long incision in the back of the head with the scalpel. He takes hold of the scalp and pulls it loose from the skull, slicing forward to the forehead to expose the cranium. Then he discards the scalpel and fires up the electric saw again.

The noise is sharp and piercing, and the burning odour clings to my nostrils. The technician saws evenly, with precision, until he is through the cranium. I pinch my nose as the technician cuts through the nerve fibres that hold the brain in place, before he lifts it out and places it on the bench in front of him.

Astrid takes her camera and moves across to take some photos. The pathologist wipes the scalpel clean and sets the brain down lengthwise on a board on the bench. 'A coup injury on the occipital lobus,' he says, pointing with the tip of the scalpel to the point at the back of the head where the indentation on the cranium had been identified. 'What does any possible contrecoup injury on the frontal lobe, or the absence of such, tell us?'

'In an accidental fall,' Spotty Neck speaks up, stepping out from behind a student colleague, 'you will often find a matching contrecoup injury on the frontal lobe. If the deceased was standing still, on the other hand, the frontal lobe will remain undamaged.'

'OK, then, are you paying attention now?' The pathologist divides the brain into two equal parts and places them in front of him with the insides facing up. He aims the tip of the knife at the paddle-shaped pattern on the frontal lobe. 'Clean and healthy,' he says. 'No sign of injury.'

Without warning, the pathologist uses the scalpel to point in my direction. 'You there, sleepyhead! What's your name again?'

'Aske,' I answer meekly, struggling to straighten my back and breathe normally through the surgical mask.

'OK, Aske. What have we learned so far today? No, I'll rephrase that: what have *you* learned today? Tell us that instead.'

I clear my throat as the group of students huddled round me all gravitate away, as if aware that this will end in yet another verbal humiliation. 'That the police were right. This is a suspicious death.'

'Oh? Why do you say that?' The pathologist sets the scalpel aside and folds his arms. Astrid also glances up from her notes and peers inquisitively at me in the crowd of students. 'Why not a diving accident? A fall and then a drowning?'

'The head injury indicates that he was standing still. That excludes a fall,' I answer, clearing my throat so that I can breathe more easily as I speak. 'And considering the circumstances and the fact that he is attached with cable ties to a woman's arm—'

'So you mean that we can exclude possibilities such as decompression sickness, acute pulmonary oedema, pneumothorax, and air embolism as cause of death, is that what you're saying?'

I nod, even though I don't entirely grasp what all that means.

The pathologist stands gazing at me for a long time without uttering a word, until in the end he says: 'Peculiar reasoning from a medical student.' He waggles his fingers in mid-air. 'Well, Aske.' The pathologist tilts his head towards one shoulder. 'Then in fact we might as well just pack up here and send Astrid to arrest that poor suicide candidate up in the intensive care unit for murder – what do you think?'

'I doubt that would help,' I whisper before I am interrupted again.

'It seems as if Aske here has watched too many episodes of *CSI* instead of reading the set texts. Like so many of the rest of you, he doesn't appreciate that it is our job to chart and document the pathological evidence we're faced with before we stitch up the bodies and wheel them back into the refrigerator. Job done. Pathologists who solve murder mysteries are something we see on TV when we go home and want to eat popcorn with the cat or our pet stick insect. This is a matter of routines and procedures. Flesh and blood. Cadavers to dissect. Death's gift to life is knowledge, Aske. Knowledge, not vague guesses and half-baked insinuations.'

Finally he stops playing with his fingers and pulls off his gloves, tossing them into the bin together with his face mask. 'Right, I'm tired of all this talking. We'll stop for lunch. Histology after the break.' He leans across to Astrid, who is packing up her camera, and mutters under his breath: 'For some reason I always get an incredible urge to eat a Toblerone after these sessions. What about you?'

CHAPTER 48

I stagger away from the autopsy suite, tearing off my coat and boots once I reach the changing room. I am already out of the door and in the corridor before the next student arrives. In the lift I take off the gloves, hairnet, trousers and face mask, rolling them into a ball that I hide in a soiled-linen cart beside the goods lift outside the ward in the renal unit.

The old man at the window is gone, although his clothes are still here and the bed is not made. I turn and venture into the bathroom, where I stand in front of the mirror. My face is ashen, my eyes pale, almost lifeless. My hair looks like a ruined bird's nest and the bristles on my chin stick out from my wan complexion like spines on a sea urchin.

I try to smile, striving to force out one of the most fundamental expressions of emotion in a human being's repertoire, but nothing comes. My facial muscles will not flex to perform the gesture. It strikes me that what I am looking at is a death mask: the cast of a face. A physical reminder of someone from the past, something I have to carry with me wherever I go.

I pick up a disposable razor from the shelf beside the basin and start to shave, laboriously scraping away the bristles from every furrow and scar on the damaged side first, and then the other. Once I'm finished, I strip off and step into the shower.

The water is cold. I stand right under the spray, closing my eyes as I soap my body, face, and even my hair in an effort to wash away the oppressive, choking odour of the autopsy room.

All the time I'm doing this, I can see Rasmus Moritzen in my mind's eye. Not the shell devoid of brain and guts as I had just seen him on the metal table down in the autopsy suite, but a combination of the man in the picture belonging to Arne Villmyr and Anniken Moritzen, and the lifeless body on the floor beside me on board the rescue helicopter. Rasmus Moritzen was murdered. Rasmus Moritzen is tied, no, attached with cable ties, to the woman I found out at the lighthouse. Bjørkang and the sergeant are still missing, and the police believe that they came out to the lighthouse after all. It means that everyone who knew or has been in contact with the woman I found in the sea is either dead or disappeared.

When I'm done, I cross over to the mirror again, where I use my index finger to draw a circle in the condensation. I give the circle eyes, nose and a line for a mouth, and lean in close to it.

'Apart from me,' I say to the face in the mirror that smiles vacantly back at me. Around the edges, tiny drops of steam are forming. 'And you.'

A perpetrator. Gunnar Ore and the crime-scene examiners were at least right about that. There is a perpetrator. Whoever emerged from the sea and took the body away is probably the person who killed Rasmus.

I lean in closer. The droplets of water on the mirror are so large that they start to trickle down the glass and over the eyes, nose and mouth on the circle face, as if they were tears. 'Was that why you took the woman without a face down into the disco, because you knew that I would never leave the place alive?' I ask the face that is rapidly disappearing in front of me. 'Is this a game? Information and manipulation? Is that what this is all about?'

I stand facing the mirror until the ventilation system has sucked all the moisture out of the room and the glass is bone dry again. 'OK, then,' I add without taking my eyes off the mirror. 'Then let's play.' I turn on my heel and leave the bathroom.

CHAPTER 49

I make for the footboard where the old man's jacket is still hanging and rummage through the pockets until I find his mobile phone. Once I have brought it over to my bed, I key in Ulf's number.

'Hi, Ulf. It's me. Your favourite patient.'

'Are you aware how much shit I've—'

'How are things in Stavanger?' I ask, breaking into the brewing storm of recriminations and insults.

'What if you took a big stick and just shoved it' – his normally so therapeutically oriented vibrato, with overtones of a Bergen accent, is working in top gear – 'right up your arse, you bloody moron! Ruth is furious that I even let myself be persuaded to take you under my wing, and—'

'Ruth? Who is Ruth?'

Ulf pauses his tirade for a moment before answering in a slightly calmer tone: 'Ruth, my live-in partner, for heaven's sake. She's a colleague I met a while ago at a conference in Drammen.'

'I thought she was called Solveig.'

'No,' Ulf says, spluttering. 'Solveig moved out the same day you headed north. Took Frida with her and went home to Bergen.'

'I'm sorry to hear that.'

'Nothing to be done about it, nothing more to be said. These things happen. Even to psychiatrists.' He fills his lungs with cigarette smoke. 'So, what's going on, Thorkild Aske? Could you please tell me? What on earth have you done?'

'What do you mean?'

'Listen ...' Once again Ulf is on the verge of exploding, but instead pockets his pride, takes a deep breath and goes on: 'A few minutes ago I ended a conversation with a Dr Weidemann of the renal unit at Tromsø University Hospital, who wanted to know something about a patient they'd admitted to the trauma reception facility last night. My very own Thorkild Aske, who they assume threw himself on purpose into the sea while chock-full of medicines. My medicines!'

'Technically speaking, the medicines were mine,' I correct him. 'Besides, my drowning attempt came to nothing.'

'What? What! But bloody hell ...' This time Ulf is unable to hold back his fury. After an all-encompassing, thundering speech in which he threatens everything from premeditated murder to compulsory psychiatric treatment on a rigid kill-or-cure regime of Paracet and Truxal, he finally cools his jets and is ready to talk again.

'Well,' he says, sounding calmer, 'what is going on?'

'I found Rasmus in the sea. He'd been murdered.'

'Murdered?!'

'According to the post-mortem he had sustained a head injury caused by a blow, and died of drowning. His body was attached with cable ties to a woman's arm, and I reckon it was the same woman that I found out at the lighthouse.'

'Post-mortem? A woman's arm? But what the ...' Ulf stops to take a deep breath. 'OK, OK, OK,' he intones between inhalations. 'We can discuss this further some other time. But what's happening to you now?'

'I've just had an enema.'

'Brilliant, brilliant,' Ulf murmurs distractedly as he lights another cigarette.

'Ulf,' I start to explain, 'I didn't think about how this would—'

'OK, not now. We'll go into that when you get home. Because you are coming home, aren't you?'

'Not yet.'

'What? You must travel home to Stavanger at once. This has spun completely out of control, and you have no business getting entangled in a criminal case. I'll phone Anniken right away and tell her that things have gone too far, and you can't help her any further. Then you can come home today.'

'That's not on,' I quickly contradict him. 'I have a police interview tomorrow that I can't get out of.'

'Police interview?' Ulf's voice drops to a softer octave. 'Why's that?'

'They found blood from one of the missing policemen during a crime-scene examination out there at the lighthouse. Presumably, they've also found something else out there that can be linked directly to one of the two policemen.'

'But what does that have to do with you?'

'The evidence connects Bjørkang and his sergeant to my whereabouts that evening at the lighthouse.'

'And?'

'Don't you hear what I'm saying? Two policemen are missing after having gone out to a lighthouse to meet a mentally unstable, suicidal, brain-injured, ex-policeman who has just been released from prison.'

'Bloody hell.'

'It will all come out this time,' I continue. 'Frei, all that shit, no matter if I'm guilty or not.' I hesitate for an instant. 'She was there again the last time I was out there. Not Frei, but the woman without a face. She was sitting down there in the disco. At least I think she was – that's why I have to go back there again. I have to see for myself, and know where the dividing line is. The line between fantasy and reality, before I step inside that room. In advance of the interview.'

'Now I understand,' Ulf says, energetically stubbing out his cigarette as he exhales, feeling disgruntled. 'Is there anything I can do to help?'

'Yes. Get me out of here.'

Ulf clears his throat. 'OK. Where are you?'

'The renal unit. They won't tell me when I'll be allowed out.'

'OK, OK,' he chants to himself in some sort of stress-releasing mantra. 'If they regard you as suicidal, psychotic or see any sign of serious mental illness, they will use a compulsory order to admit you to an acute psychiatric ward for observation.'

'How long would I have to stay there?'

'Up to ten days,' Ulf answers.

'I don't have time for that.'

'I'm afraid you don't have any choice.'

'Help me, Ulf,' I beg him as I hear footfalls in the corridor, heading straight for my room. 'Help me, for God's sake.'

'It's too late, Thorkild,' Ulf sighs. 'This time you've got yourself too far tangled up in the system for me to help. Take those ten days, Thorkild – I think you need them, considering all you've been through. Maybe I can do something to shorten the time; that's not impossible if you just—'

'Time's already up, Ulf,' I whisper as the door opens and a doctor pokes his head in.

'Thorkild!' I hear Ulf yell just as I move to disconnect the call.

'Yes?' I put the phone to my ear again. 'What is it?'

'Don't fool around now.'

'Fool around?'

Chapter 50

'You're back,' the doctor remarks as he enters. He has a cup of coffee in one hand and a brown envelope in the other, resting it on his chin as if intending to use it as a fan. Grey-haired and about my own age, he speaks an eastern Norwegian dialect.

'What do you mean?'

'We were here a moment ago.'

'I was in the toilet,' I answer. 'Stomach problems.'

The doctor looks down his nose as I try to summon the necessary facial gymnastics to make it both embarrassing and believable at one and the same time. The smell of the autopsy room lingers in the air, as if already impregnated in the bedclothes and walls.

'Where are your clothes?'

'They were cut up downstairs in trauma reception and are in a bag here somewhere.'

'Have you anybody who can bring a change of clothes for you?'

'No.'

The doctor casts his eyes down at his shoes. 'OK,' he says after a pause. 'I'll see if we can organise something for you from our lost property department.' He sits down on the chair beside the bed. 'How are you feeling today?'

'Old,' I reply laconically.

'Dr Weidemann.' He offers me a sinewy, tanned hand with perfect fingernails.

'Thorkild Aske,' I reply, sinking back on to the bed.

My body feels suddenly heavy, as if exhausted and drained of energy. 'So,' I continue, now that the pleasantries are done with. 'I'm cured, then? That was quick.'

'Well,' Dr Weidemann says, with a sigh, and starts over again. 'We're transferring you to the psychiatric unit at Åsgård for observation with the intention of—'

'No, thanks,' is my response, and I smile as broadly as my facial muscles will allow. 'Just dropped in to see what condition my condition was in. Need to push off now.'

'We have the law on our side.'

'What law?'

'The reference is paragraph 3.2 of the law on mental health care and insists on compulsory observation for up to ten days.'

'Why?'

He gives me a resigned look, and only now do I notice that his eyes are bloodshot. 'Considering what has taken place, we're afraid you might be a danger to yourself.'

'Oh?'

'Here.' He passes me the envelope, and I weigh it in the palm of my hand, fixing my eyes on the man by my side, crumpling so far into the chair that it looks as if he might bend in two. 'Can I read it?'

'Er, yes, yes of course. Inside, you'll find mention of the paragraph about compulsory admission to an acute psychiatric ward where you can find the legal authority.'

Outside, darkness is falling. I can hear the wind whistling through gaps in the window frame just as I catch sight of two men who have appeared in the doorway. One is young, with short hair, and the other older, with a terrible haircut. Both appear to have lifetime subscriptions at a weight-lifting club and doctorates in physical confrontation.

'Who are they?'

'They're here to assist.'

'You?'

Weidemann shakes his head. 'In case you need it.'

'Assist with what?'

'Anything.'

'When am I leaving?'

'Now, or as soon as you feel fit enough to make the move.'

'Can I refuse?'

'Yes.'

'Fine, then I refuse.'

'But ...' Weidemann raises his hand again, like a white flag between us. 'But ...' he repeats with renewed determination. 'Then the police will drive you there.'

'Fantastic. Then I'll go under my own steam.'

'Taxi or ambulance? Jørn and Jørgen here will escort you.'

'I don't have any money.'

'We'll arrange that.'

I spread out my hands in consternation. 'Doesn't look as if I have any choice.'

'Fine, then.' Dr Weidemann drags himself out of the chair and shuffles over to the door while the two care assistants step into the room and stand smiling in front of me. Not the warm, welcoming smiles you receive from someone you know, but rather two cold invitations, as in an *even now, here we stand side by side with you, a trouserless ex-policeman with nothing to lose, and still we're smiling* type of smile.

I clasp my hands reverently over the brown envelope and return the smiles, acting as friendly as I can possibly contrive, in view of the circumstances.

'Hey!' I call out after Dr Weidemann as he makes for the door.

'What?' He stops and wheels round.

'Can you give me half an hour?' Laying my hand gingerly on my stomach, I pull a grimace. 'Doesn't feel as if I'm finished in the toilet after that colonic irrigation.'

'Of course,' he answers. 'In the meantime, I'll see if I can find any fresh clothes for you.'

The two men stand just inside the door as I haul myself out of bed and hobble bent-backed over the linoleum and into the bathroom, where I close the door behind me and turn on the basin tap full blast.

I sit on the toilet-seat lid and take out the old man's mobile phone that I have kept concealed in my hand. Time to take responsibility and show some personal initiative, as Gunnar Ore used to say when our section's football team was being thrashed by the legal firm from across the street.

'Thorkild?' Liz gasps breathlessly when she hears my voice. 'Oh my goodness, I've been so scared—'

'Afterwards, Sis,' I whisper into the handset as quietly as I can. 'I need your help. Now, right away.'

CHAPTER 51

The goods lift in the renal unit is either the most wretched place that anyone could hang out in, or else Dr Weidemann has a vile sense of medical humour that I can't fathom.

I'm dressed in a pair of jogging bottoms with dark blue and green stripes down the side, the kind you only see on drug addicts and teenagers in the Eastern bloc. The top I'm wearing is a thick, greyish white, woollen turtleneck sweater that keeps me constantly itchy and is too small for me.

My footwear is a pair of washed-out, well-worn trainers, out of date in every way. My own wet, sliced-up clothing has been handed to me in a yellow plastic bag from the Prix supermarket chain. I feel like a travelling poster boy from some obscure anti-heroin campaign in northern Norway, as I walk with my beaming bodyguards, Jørn and Jørgen, along the renal unit corridor towards the lift. The intention is that they will escort me to the main entrance and on by taxi to Åsgård, Tromsø University Hospital's psychiatric unit located on the western side of Tromsøya Island.

It appears my pair of travelling companions delight in the awkward silence between us in the lift – no exchange of glances, no words, just cold smiles, as if they're taking part in a competition that means whoever speaks first, or shows any other trait of humanity, has lost.

'Hey, you!' I lean towards Jørn or Jørgen on my right side. He doesn't answer, but looks in my direction, as if expecting me to continue.

'Do you know what's red and says blob blob?' I ask in an undertone without taking my eyes off the lift door in front of us. The lift stops and the door opens, but no one enters. 'No?' The door closes and the lift continues its descent. 'Well, don't let me keep you in suspense any longer than necessary. It's actually a ...' I delay for a moment before pressing on: 'Last chance. You get one more chance to guess what it is.'

Jørn or Jørgen on my right side still makes no response, and Jørn or Jørgen on my left doesn't bat an eyelid, instead simply concentrating intently on his statue impression as he listens. 'OK, here it comes ... a red blob blob!' I exclaim, nudging his shoulder as I guffaw and clap my hands against the carrier bag's soft plastic.

The lift stops again. Jørn and Jørgen indicate that we have reached the right level, and that I should move. We advance into a bright department and set off along a wide corridor towards a huge space that I realise must be the exit area, even though I can't see any exit doors for the crowds of people.

'And you?' I prod the other Jørn or Jørgen's shoulder, the one on my left-hand side. 'Do you know what's red and says blob blob?'

He doesn't answer either, and merely steers his way purposefully through the wheelchair users, pregnant teen-age mothers, patients with plasters and bandages, and their relatives.

'Don't you know?' I say, nodding as we pass two men standing with a ladder, busy changing a fluorescent light tube on the ceiling. 'I just told you, you know,' I continue. 'But that was wrong. That's what's so funny. It's not a red blob blob. Not this time. It's a cranberry with an outboard motor! Get it? It changes every time. So you can't win.' I jab his shoulder again. 'You just can't win.'

'Hey,' Jørn or Jørgen on my right side whispers. 'What if we just calm down a bit now? OK?' On the other side of the ladder and the men changing the fluorescent tube, I spy a row of red chairs, and behind these a room divider half a metre high with plastic plants and clay pebbles. Beyond the plastic plants lies a spacious enclosed area where patients and relatives are huddled around tables, some reading newspapers, while others are eating or chatting on mobile phones.

'Yes, OK, of course,' I answer. 'I just wanted to explain how the same joke—'

'Yes, we get it.' Jørn or Jørgen to my left raises his voice a notch. 'You can't win because it changes every time.'

'Exactly,' I say, laughing, and poke him, light as a feather, on the shoulder again. 'You just can't win. No matter how many times you try.' I lean towards him once more, as if to give him yet another harmless little prod, but instead shove all my weight against his shoulder and ram him into the group of seats occupied by patients and relatives, so that he trips over the room divider with plastic plants and clay pebbles to land in amongst the dining tables on the other side. Then I slap the yellow plastic bag full of my wet, sliced-up clothes straight into Jørn or Jørgen's face, the one on my right-hand side this time. Losing shoulder contact as he takes a step to one side, he collides with the ladder that one of the men has just begun to climb, holding a replacement fluorescent tube in his hand. The result is that the man drops the light tube when he has to grab the ladder and it smashes on the floor.

As for me, I break into a run.

I turn to the left, past the seating area, and hare off after a group of people in outdoor clothing just beside the entrance to a pharmacy – they are making their way towards a sloping corridor that seems to lead outside. Unfortunately my hunch turns out to be wrong, and the corridor simply runs down into another, narrower passage.

I turn tail abruptly, about to retreat, when the younger Jørn or Jorgen suddenly cannons into me from the side, so that we both collapse on to the floor and roll up against a cardboard box full of items on special offer, on display at the pharmacy entrance.

My arms flail in an effort to recover my footing and I throw a bottle of sun cream, reduced to half price, straight at the younger Jørn or Jørgen, hitting him smack in the face, just as he is about to jump me. I scramble to my feet again and set off towards an information desk, where I spot a large sign and an arrow showing the way out.

All at once I spot the older Jørn or Jørgen, standing right between a pillar and me. The exit is behind him. He holds his hands out at hip level, snorting like a quick-tempered bull, as he glares furiously at me. The sight of him makes me slow down, calling to mind a duel scene from some old Western film in which the townspeople – in this case a glorious mixture of battle-scarred patients and cancer sufferers desperate for a smoke – run in panic in every direction until there are only the two duellists left in a main street blowing with dust. The problem is just that neither of us has a weapon, if neither of us is keen to grab a walking frame from someone in the crowd of spectators or a tube of wart ointment from the pharmacy.

I pick up speed, not towards the waiting bull in front of me, but veering instead to the right, through the groups of people seated at tables, back to the corridor we had come from. Out of the corner of my eye, I see that the older Jørn or Jørgen is sprinting through the crowd to cut me off.

I force my way past the room divider and race towards the lifts in the forlorn hope that one of them is open and available to get me out of here. Pivoting round for a second shows me that both Jørn and Jørgen are only a few metres behind. I'm never going to manage to close the doors behind me before they catch up. I fling open a glass door at one side of the lift, then slam it shut again and keep hold of it for a moment as my lungs whistle and the sweat runs down my face.

On the other side of the glass, Jørn and Jørgen push against the door, both of them panting and puffing. Behind me, the buzz from a group of people gathered round a counter falls silent.

'Can I help you?' asks a young woman in white who leans over the counter.

I have neither breath nor time to answer. Instead, I let go of the door and set off along the gable wall just as the glass door behind me springs open.

Desperately, I unlatch an identical door at the other end of the wall and am back in the corridor, only this time on the other side of the lifts. I pass the first glass door again as the younger Jørn or Jørgen opens it. I swerve out and away from him as he launches himself at me, and I feel his fingertips snatch at my jogging trousers and fail to gain a grip.

'Stop, for Christ's sake,' gasps the older Jørn or Jørgen at my back, while I tighten my hold on the yellow carrier bag and race onwards in the direction of the exit yet again.

The cold air smacks me in the face, slicing straight through the woollen sweater, but I don't slacken my pace. The taxis are parked on my right, three in a row along the kerb. In the distance I hear the rumble of heavy machinery from a closed-off area beyond the entrance, where some sort of extension is being built.

Then I catch sight of the car.

I spit the taste of blood out of my mouth. My diaphragm's on fire, so are my cheeks, but I daren't take a second's respite, because the very next minute the door bangs open behind me and I don't need to turn round to know that it's the two care assistants hard on my heels, determined to show off their skills in welfare treatment such as tripping, headlocks and strangleholds.

'Wait, for fuck's sake!' they yell in chorus as I move past the taxis towards the elderly Ford Mondeo parked in front of them with its engine running.

'Drive, Liz!' I scream, while I wrench the door open on the passenger side and leap inside.

'Wh … what?' my sister wheezes, staring at me in alarm. There's a thick band of beaded perspiration emblazoned on her forehead and the hairs above her top lip, as if she is the one who has been running for her life.

'Drive, for fuck's sake!' I repeat, and slam the car door shut at the very same moment that Liz floors the accelerator.

CHAPTER 52

The car hurtles down through the Tromsøysund tunnel that takes us from Breivika on Tromsøya Island across to Tomasjorda on the mainland. 'I've been so worried for you,' Liz laments, perched so far forward in the driving seat that her breath mists the windscreen, and clutching the steering wheel for grim death. 'I've hardly slept since we last talked. I thought you had—'

'Don't give it a thought, Sis.'

She turns towards me. 'What happened?' Her eyes scan my get-up. 'Why are you dressed so weirdly?'

'I fell into the sea,' I answer, and fish my mobile phone from the plastic bag stuffed full of my clothes to plug it into the charger cable hanging from the dashboard cigarette lighter. 'They chopped my clothes up to bits in Accident and Emergency.'

'B ... but ...'

'Just drive.'

Liz reveals a row of crooked yellowed teeth and unhealthy gums. 'I don't get it—'

'There's not so much to get, Liz. Things happen, don't they? Things we don't always want to talk about.'

Liz pouts crossly and asks no more questions. She has learned to recognise when it works best to keep quiet. Her servile, battered survival instinct functions on autopilot in such situations, thanks to the many lessons in violence taught by her creep of a husband through the years.

'Why do we have to hurry?' she asks submissively when she finally dares to open her mouth again. 'Are you in more trouble?'

'No,' I answer, drawing breath. 'I just didn't have time to go where they wanted me to.'

'I remember Dad.' Liz flashes a smile once we have emerged on the other side of the undersea tunnel. 'When they brought him home – we were only little at the time.'

'What do you mean?'

'After he'd been in Kleppspitalin, the psychiatric hospital. It was a few years before he and Mum separated. You were so small then – you probably don't remember it.'

'No,' I answer grouchily. I'm still feeling giddy and on edge after all the running. My body is shaky and longing for rest, nourishment and anti-anxiety medication.

'Are you sure?' she chuckles. 'Don't you recollect how—'

'Didn't you hear what I said?' I pounce on her in a burst of fury that makes the aches in my face explode. 'I said I didn't remember!'

'No, no, no!' Liz whines. 'You don't remember, then, you don't remember!' The flare-up makes her jerk the steering wheel, and the car swerves from side to side before she eventually regains control of it and herself.

'Sorry, Thorkild. Sorry.' She tightens her grip and leans closer to the windscreen, as if to demonstrate that now she will really exert herself and prove to me and the rest of the world that Elizabeth too can drive cars and not merely bake cakes and feel sorry for herself.

I well remember when they came home with my father after he had left to regain his self-control following bad times at home with us. They used to put him in a chair at the kitchen table, where he sat for hours just staring vacantly out at the rain and sea. It was as though he had just returned from an alien planet and needed time to acclimatise again. Eventually I came to terms with the fact that he had some kind of illness inside that made him rot outwards, corroding everything that he might have been and replacing it with this pathetic creature who roamed around weeping for Iceland, for a few strips of moss or a bug in a river in a gravel wasteland under a volcano.

'Where are we going?' Liz asks, once she has calmed down.

'I just need to go back to Skjellvik again to sort some things and pick up the hire car before leaving for an interview tomorrow.'

'Interview? Another one?'

'Gunnar Ore is in town.'

'Gunnar? What's he doing here?'

'I'm still not entirely certain,' I reply solemnly as Liz runs the tip of her tongue over her dry lips. The sky is almost completely cloudless, the air is colder, and there are brittle sheets of ice on the small puddles that dot the pavements. I check my mobile and find enough juice to make it useable, so I switch it on and call Anniken Moritzen.

'Where are you?' she asks.

'Still in the north.'

'They want us to come up there to look at him. I'm travelling up tomorrow.' She hesitates slightly before continuing: 'Could you meet me, when I have to go in and ... see him?'

I stare at the outline of my own face in the side window. My hand instinctively reaches out to the indentations in my complexion, and my fingertips trace the lines and scars until they find the pressure points. I dig into the damaged tissue on my cheek.

'What is it, Thorkild?' Anniken probes when I fail to answer.

'Rasmus was murdered,' I whisper.

'Wh ... what?'

'He didn't die in an accident, Anniken. Someone killed him.'

'I ...' She starts to speak, but stops, bites back the words, and falls completely silent. Liz steals a concerned glance at me from the driving seat without saying anything either.

I relax the fingertip pressure on my cheek as my gaze travels over the landscape and sea farther down where a plastic boat is berthed, tugging listlessly at its moorings on the shore. 'Could he have met someone up here – a woman?'

'I've already answered that.'

'I know,' I continue. 'But would he have told you if he had met someone?'

'Rasmus didn't have a girlfriend,' Anniken Moritzen replies. 'If he had, it wouldn't have been a *she*. He was gay,' she adds, 'like his father.'

'So he didn't know her,' I comment.

'What?'

'I'm trying to tie up these loose threads. To see how it all hangs together.' I use my free hand to rub the damaged tissue on my cheek. The gnawing in my belly is worsening. 'Did Rasmus use drugs?' I ask, gritting my teeth.

'What?'

'He couldn't have been involved in anything illegal?'

'Stop.' I can hear her struggling to retain self-control. 'I want you to stop talking,' she says before her voice fades away entirely. All I can hear is a continual stream of muffled panting. 'Can't you just be there,' she implores. 'Be there when I arrive?'

'OK,' I answer, breathing out noisily. 'Phone me when you land. By the way, is Arne coming with you?'

'No.'

'Does he know?'

'Yes, he knows.'

'Has he said anything about—'

'No. He doesn't want to talk to you. Not yet.'

'OK, Anniken, I'll wait up here until you come.'

'Goodbye, Thorkild.'

'My God,' Liz says under her breath after I have hung up. She has been totally silent during the phone conversation. 'What on earth have you got yourself mixed up in?'

I recline my seat to the maximum and turn my face to the sea. 'At the interview tomorrow afternoon, Ore, the boys at Police Headquarters and the Chief Interviewer from Kripos, are all going to tell me they have found proof that Bjørkang and his sergeant came to the lighthouse that evening when I was there.'

I hold my breath, watching the treetops outside, until finally I exhale again and continue: 'They're going to present me with a scenario in which a pill-abusing ex-policeman commits a murder before ultimately, filled with remorse and self-loathing, choosing to jump in the sea when the pressure mounts. Then they will ask me to give my version. They are going to sit there, quiet as mice, as I tell them about the woman I found in the sea, about the man who came up and snatched her away, about the party in the disco, about all of it. In the end they will tell me that they don't believe me, and earnestly urge me to give them the right story. The truth.'

'How do you know that?'

'Because, Liz,' I turn to face her, 'it's what I would have done. With the crime-scene finds and my own explanation, which even I find hard to believe, they consider this to be a solid case. They may even feel sure enough to make an arrest regardless of how the interview goes, if they have enough confidence that I will eventually break, even without a body.'

'What about Rasmus and the woman you found? You weren't even here when they were murdered.'

'They will end up being treated as separate cases, and they will remain unsolved as long as she doesn't turn up again – something she is never going to do,' I add. I turn back to the side window where tiny snowflakes have started to dance between the trees. 'And then you have the two policemen.'

'What about them?'

I let the question hang in the air between us while I try to conjure the faces of Bjørkang and Arnt the way I remember them from our meeting in the local police station. I am unable to summon them; they remain two grey, shapeless shadows. 'I have to find them before this interview. It's the only way out …'

CHAPTER 53

Liz parks in the car park outside the Skjellviktun Residential Centre. It will soon be half-past eight in the evening, and it's dark outside. The air is cold and raw when we leave the car and walk to the apartments.

'No one can find out we are here,' I whisper to Liz as we cut across the gravel to the apartment block entrance where Harvey has given me a place. My eyes dart in the direction of the residential centre entrance when we pass the door, for fear that Siv, the nurse, will emerge with a police escort to take me to the ward that waits for me at Åsgård psychiatric unit. The corridor inside is deserted, and everywhere around us is silent – the whole place seems to have gone into hibernation.

I produce the front-door key from the bag of ruined clothes and open the door without a sound, pushing Liz inside before closing and locking it behind us.

'No light,' I whisper, as soon as we are inside.

The truffle cake is still in the cake container on the kitchen worktop, and Merethe's crystals are scattered all over the coffee table and floor. Liz settles on the sofa while I find some clean clothes from my travel bag and take them with me into the bathroom.

Pausing in front of the mirror in the gloom, I study my face in the glass. So worn out, so grey, almost merging into the darkness. 'There's no one else in there, is there?'

I whisper huskily to the shadow in the mirror. No more Thorkilds to be summoned when things get too difficult or when the opportunity presents itself. This mottled, dismal bag of bones is all that is left. *Good God.*

At last I find a flannel and wash my face and armpits before stripping off the outfit from the lost property department at Tromsø University Hospital and replacing it with my own clothes.

'Everything all right?' Liz asks when I creep back into the living room. The coffee machine on the kitchen counter is rumbling and the lid of the cake container is lying open beside it. Liz sits there in the dark with a large slice of truffle cake in her hands. The streetlamp outside in the car park and the light from the main entrance send yellow glimmers in through the windows and curtains so that we can see slightly. 'I put out a piece for you as well.' Liz pushes a paper plate in my direction.

'No thanks – just you eat. I have to make a phone call.'

I am about to sit down when a light appears outside. Through the gap in the half-open curtains, we see a police car glide over the brow of the hill and drive slowly past, heading towards Skjellvik town centre.

'Is it you they're looking for?' Liz asks with her mouth full of cake.

Nodding, I close the curtains. 'They've probably sent a car from Skjervøy.'

Liz stops eating and looks at me.

'Relax,' I whisper, taking out my mobile. 'They won't find us here.'

I sit down in the chair and call Gunnar Ore, who answers before the first ringtone has ceased.

'On the run?' he asks acidly. 'Good move, Aske. Really smart.'

'Yes, I'm on my way over the mountains to the Swedish border,' I reply. 'And then perpetual asylum on the Andalusian steppes dressed as a shepherd or a matador, depending on how things turn out. It's possible, Gunnar. It's possible.'

'You idiot,' he responds hoarsely. 'The police are searching for you. You're aware that the hospital sends an alert to the police as soon as a psychiatric patient escapes?'

'Let them search.'

'Then you must also know that the police have been armed since the time you left?'

'What?'

'The boys drive around with rounds of live ammunition these days, Aske. And in your case they're also searching for a suspected police killer on the run. Do you need any more details to understand how this is all going to end?'

'I just wanted to say that I'll come in tomorrow, to that interview,' I say.

'Too late,' my former boss counters. I am well aware that the police in Tromsø must have called on Gunnar Ore for assistance as soon as they knew who had fled from the hospital. Nevertheless, I was obliged to talk to him, even though the thought of ending up as a trophy for some recently qualified, trigger-happy police officer scared the wits out of me. I have nobody else to turn to.

'Where are you?' Gunnar continues when I do not say anything.

'With Liz,' I answer.

'Your sister? So she was the one who picked you up outside the hospital. OK, now I get it. Why? Can I ask you that? Why did you run off?'

'The interview,' I reply as Liz returns from the kitchen counter with a steaming mug of coffee for me, and a fresh slice of cake for herself. 'I can't turn up for that interview until I know.'

'Know what?'

'Where I stand. What this is actually all about.'

'Meaning you're back in Skjellvik?' Gunnar Ore concludes.

'I'm coming in for interview,' I say under my breath. 'I promise. Give me until tomorrow, for old times' sake. Can you do that?'

'You played that card four years ago, Thorkild,' he answers. 'But I'll forget this conversation until three o'clock tomorrow. For old times' sake, as you call it.'

'Thanks.'

'Oh, fuck off. In the meantime you'd better hope that those policemen turn up by three o'clock tomorrow. If not—'

'This is going to end fucking badly?'

'It already has done,' he answers, then hangs up.

I put the phone on the table and raise the coffee to my mouth, but have to give up halfway and use both hands to hold on to it to avoid dropping the mug.

'What is it?' Liz asks, sounding worried. Her facial expression is hidden even though she is sitting just beside me. The aroma of rich truffle cream and coffee wafts towards me as she speaks. 'Are you unwell?'

'Forget it,' I say, leaning further back in the chair to dodge the smell. 'It's just my body trying to tell me that it's long past time for my medication.'

'I have some Paralgin Forte in my handbag,' Liz tells me. 'They belong to Arvid. For his back pains. I always carry them with me when we go out, in case he suddenly needs them.'

I take the blister pack she offers me and press out a handful of pink pills that I toss into my mouth to stifle the screams deep inside me.

'What will we do now?' Liz asks when I finally feel ready to tackle the mug of coffee. Another car passes outside, but this one also drives on. The sky is blue-black on the horizon out at the lighthouse.

'We'll stay here overnight and tomorrow you can drive home again.'

'And you?'

'I'll follow in the hire car. First I just need to pay another visit to the lighthouse.' Reclining in the chair, I close my eyes and massage my cheek with my fingertips. 'Oh my God,' I groan. 'Where the hell can they be?'

'The two policemen?'

'And the woman I found in the sea. It all started with her, that's the only thing I'm sure of. She was first to disappear, almost a month ago. Who can have vanished for so long without anyone reporting her missing?'

'Someone who's not from here?' Liz asks.

'Yes.' I open my eyes and sit up straight again, then grab the blister pack and pinch out two more tablets. 'But not only that,' I go on after swallowing them down with a swig of coffee. 'She must be someone who was meant to travel here, and who still isn't expected back wherever she came from. Don't you agree?'

'A tourist?'

I look at Liz. She has moved to the very end of the sofa and is sitting beside me with her hands on her lap. 'Yes, OK. But who comes here as a tourist in late autumn?' I ask.

'Maybe she came here to work. From another country?'

'And who doesn't phone home to her husband, boyfriend, children or family for more than a month?'

'Maybe she has nobody,' Liz suggests.

'Nobody at all?' I drink some more coffee. 'Young, alone, come from another place and have nobody who misses you. What sort of woman is that?'

'A prostitute?'

'What?'

'They come to Tromsø too, book into hotels and bedsits for a few months at a time, before travelling on,' Liz informs me. She is talking faster now that she sees I am interested. 'Yes, I've read that in the newspapers—'

'Great, Liz,' I say when I feel the pills finally beginning to work. 'Really great. Well, maybe you've hit on something there. A young woman, a prostitute. But what would she have to do with Rasmus?'

'He took her with him out to the lighthouse to—'

'No.' I shake my head. 'Rasmus was gay. Besides, he was murdered. I think he found her by chance when he was out diving. And then what happened?'

'I've no idea,' Liz answers.

'Yes, you do,' I press her. 'He called the police.'

'Yes,' Liz exclaims, kneading her fingers and looking at me, wide eyes glinting in the gloom. 'Yes, that's what he did.'

'OK, he phoned the police, to tell them what he'd found. And then, let's complete the scenario: Bjørkang and Arnt

316

came to Rasmus and killed him because no one could know what he had found? Is that what you're saying?'

Liz shakes her head. 'I ... I,' she stutters before her eyes stray to the cake container. 'I don't know, Thorkild,' she says in the end.

'No, I suppose not. That's where it all goes astray. I mean – why didn't they come back afterwards and keep up whatever game they were playing as soon as they had picked up the body? Why go into hiding?'

Liz gives me a strange look: smiling, her eyes gleaming, as if brimming with tears.

'What is it?'

'I like to see you when you're like this.'

'Like what?'

'Like now, when we're talking to each other, just the two of us. Do you remember Dad and his friends used to sit round the kitchen table discussing what had to be done to defend Iceland against industry?'

I lean back in the chair again and turn to face Liz. 'Last time I saw him was in the early nineties,' I say in a murmur. 'In a little village called Reyðarfjörður where the state-owned energy company, Landsvirkjun, was planning to build an aluminium plant. I had just learned that I had gained entry to Police College.' I shut my eyes and compress my lips as I cast my mind back to that grey, rain-soaked village. 'That white hair and beard of his, my God,' I say, laughing. 'What did he look like?'

'He still has them.' Liz chuckles again, blinking, and keeps her eyes closed for a few seconds, as if trying to picture him.

'The aluminium plant was going to be run by an American firm. Dad and his crowd of eco-warriors had gone out there to see if they could sabotage the work. It was in the middle of summer, and we sat there drinking wine around a bonfire while Dad talked about our country.'

'I wish I could have been there.' Liz squeezes her hands against her chest.

'I remember sitting there waiting for the right moment, when I could tell him that I was going to be a policeman in Norway. Dad had pulled up a fistful of heather and soil, tears were running down his cheeks and into his beard, he was weeping buckets on the other side of the fire as his fingers grasped this little clod of Iceland that he clutched to his chest. *How long will they be allowed to rape the natural environment of our beloved Iceland?*'

'So you didn't tell him?'

'No,' I answer. 'We just sat there, all of us, listening to Dad talking about heritage and homeland. Then we drained our wine glasses and headed down to the fjord, where we rolled the container and barracks the first engineers had set up out into the sea. When we got back again, I packed up my tent and went home to Norway. One month later I started at Police College.'

A lengthy pause ensues. 'Can you tell me something about Frei?' Liz eventually whispers. Her voice is subdued, the way she must speak when she is alone with a husband who beats her. 'Who she was, why you are so … hung up on her?'

'I slipped up,' I tell her, still with my eyes closed. 'I knew I loved her that last evening in the car, the game was over, and there was nothing I could do to change it.' I take a deep breath and exhale. 'The next second she was gone.'

'Is that why you want to die?' Liz puts a wary hand on mine. Her touch radiates warmth along my arm, all the way up to my face.

'I've crammed so much into that moment,' I go on in a whisper. 'But who can live like that? In a single split second?'

'You mustn't do it again,' Liz howls, launching herself at me so that we're both pushed back into the chair and it creaks beneath the strain. 'Listen to me, Thorkild. You mustn't do it again.'

'Good God,' I gasp, struggling to haul her off me.

'Promise me that.' Her face is right in front of mine, and she is holding me in a vice-like grip.

'I can't—'

'Yes you can,' she pleads, refusing to let go. 'If you promise me, then you can't do it. You'd never have done it. I know that. I know ...'

I make another effort to extricate myself, but Liz refuses to budge and clings to me as if I were a lifeboat. In the end I capitulate. 'OK, Sis,' I say, sinking into the seat. 'I promise ...'

WEDNESDAY

CHAPTER 54

Next day I wake in the chair, roused by Liz washing up in the kitchen. She has cleared Merethe's stones from the floor and collected them in a pile on the coffee table. I switch on my mobile and see that I have heaps of alerts and messages on my voicemail. Including a number of rancorous comments from Ulf, who has obviously also learned that his favourite patient is on the run. I switch off the mobile again and go to Liz for a cup of freshly brewed coffee.

'The police car drove back last night, after you'd fallen asleep. I haven't seen it since,' she says, with a smile. Her grey and blonde curls stick out in all directions, and more than anything she looks like one of Ivo Caprino's trolls as she stands humming to herself with the washing-up brush in one hand and one of the paper plates in the other.

'Brilliant,' I say, giving her a brief hug before taking the coffee over to the chair and pressing out two pink pills from the blister pack she gave me yesterday. 'Then we can drive down to my hire car at the boathouses when we're ready.'

'How were you thinking of getting out to the lighthouse?' Liz asks, as she pulls the plug out of the sink.

'I know someone I can ask,' I say, massaging my neck muscles. 'Afterwards I'll go to Tromsø for the interview.'

'And then, what's going to happen after that?'

'Well, that's not up to me, Liz,' I tell her, forcing a lopsided smile. 'Is it?'

'But if you find those two policemen, then they can't very well—'

'Yes, I know,' I say, sighing, and swallow the pills with a mouthful of coffee.

'I feel sure you're going to find them,' she says, gazing at me with the washing-up brush in her hand and a cheerful, almost saintly, smile on her lips. 'And then everything will be fine again. I believe so, in fact. I felt it in my bones when I woke this morning. That this is going to be a good day.'

'Oh?' I grunt, shaking my head. 'Well, if you say so, then it must be true ...'

Once we have finished in the kitchen, we bring my belongings and drive down to the car park beside the boathouses in Skjellvik. Liz waits in the car while I dash across to the hire car to start it, darting continual watchful glances at the top of the hill, in case of spying a car with flashing blue lights come rolling over the summit.

'Wish me luck, Sis,' I yell through the open door, signalling to her that she should drive off.

Liz sends me a meaningful look and leans across the passenger seat. 'Remember, you promised,' she says, her mouth trembling.

'Yes, Liz, I promised.'

Next minute, the Mondeo fires up and she backs out on to the road at full throttle, moving to the top of the bay without taking time either to close the car window or turn off the indicators.

Just as her vehicle reaches the village, I spot Johannes in the rear-view mirror, heading straight for the cars. He is wearing a suit and his thin grey hair is slicked back. The wind takes hold of it on the side facing the sea, making it stand up straight, blowing and dancing above his head.

'Well,' he says, greeting me as I emerge from the car. 'Are you still here?'

'Yes, I'm not quite done here yet.'

'I heard the police were here looking for you.'

'Yes, they probably were. They're always so helpful, aren't they?'

Johannes utters a grunt, and runs a furrowed hand through his hair. He tries to tidy it, but the wind takes hold again as soon as his hand is removed. He is wearing black-and-grey-striped suit trousers that are several centimetres too short and have gone a few too many rounds in the washing machine; the colours are faded and the fabric bobbled. His shoes at least are black, with thick soles. On his upper body he has a crumpled dress shirt with a greyish V-neck sweater on top: it too is unravelling at the seams. On top of all that he has a tailored dark coat reaching to his knees.

'Harvey told me it was a friend of yours who found Bjørkang and Arnt's boat,' I say.

'Yes, it was on the foreshore at Reinøya,' Johannes answers. 'The boat was in good condition – no sign of it having capsized in the storm. The only thing missing was the navigation equipment.'

'What do you mean?'

'Ripped out,' Johannes says. 'Only the leads left behind.'

'Rasmus's GPS was also missing,' I inform him. 'Sounds like someone doesn't want us to know where they were going, or had been.'

Johannes nods thoughtfully.

'I have to go out to the lighthouse,' I tell him. 'Can you help me?'

'Well,' he says, poking the gravel with the toe of his shoe, 'I'm on my way to Andor and Josefine's funeral.' He is struggling to find somewhere to put his hands, since they don't seem able to fit into his trouser pockets.

'I can wait.'

He sighs heavily. 'OK, then. I expect I can run you across after the burial.'

'Hop in, and I'll drive you.'

Johannes slams the car door and fastens his seatbelt. As soon as I start the engine, he pulls a packet of rolling tobacco from inside his coat and puts it on his knee. In a matter of seconds he has produced a straight, perfectly rolled cigarette that he pops between his teeth and lights with a match.

'By the way, do you think Merethe will be there?' I quiz him as I reverse out of the car park and set off up the hill towards the village. 'At the funeral?'

With the cigarette held between his lips, Johannes folds his hands devoutly, twiddling his rough thumbs. 'I don't know,' he says as a stream of tobacco smoke pours out through his nostrils and mouth. 'I honestly don't know.'

CHAPTER 55

I feel as if I've just rolled in on a burning wheel when I emerge with Johannes from the car outside the church and encounter the cortege of bald-headed, grey- or receding-haired, wrinkled, stooped and hobbling men and women in the late autumn of their years.

In the church nave, we are greeted by curious looks from people who have already found their places on the hard blue wooden seats. Johannes points at one of the back rows where two elderly women, both in fox furs and fancy hats, are sitting in splendid isolation on a bench.

As I move to sit down, I catch sight of Harvey seated with his son several rows farther forward on the opposite side of the aisle. Merethe is not with them.

I turn towards the exit when a minibus stops outside the door with a hollow thud. Two care assistants jump out and skirt round the vehicle with the driver, who lowers the ramp and opens the back doors.

'Merethe,' I gasp as I glimpse the face of the woman wheeled out of the vehicle and up the ramp outside the church entrance.

'Where?' Johannes turns to me just as one of the women on our bench clutches the tail of her fox fur, indiscreetly nudging her companion and pointing at the entrance.

The faint buzz grows inside the church, and several people draw together, turning round to glance at the woman who is trundled in. 'My God, what does she look like?'

'Dreadful.' Johannes turns his eyes to the front, towards the altar and the two coffins. A child starts screaming somewhere in the crowd, loud, racking sobs that the parents try to stifle by alternately pleading and threatening him into silence.

The metal contraption jutting from both sides of Merethe's head comprises three screws on either side, one on her upper and one on her lower jaw, and one that locks these together. The corners of her mouth are swathed in white bandages.

Her hair is elegantly piled up, and she is wearing a flattering black trouser suit with a white blouse underneath. The care assistants push her across to the bench where Harvey and the little boy are waiting.

'I have to speak to her,' I whisper as the two care assistants lock the wheels on her wheelchair. They depart and close the church door behind them.

'After the burial,' Johannes answers.

The nave eventually fills with wheelchairs, walking frames and other gear on both sides, so that those who want to sit near the front have to pick their way through this geriatric labyrinth of rigid plastic and metal. Johannes takes out a transparent bread bag filled with sugary, red-striped mothballs. He opens the bag and offers it to me: 'Humbug?'

'Sorry?'

'Mint humbugs?' Johannes repeats in a louder voice, rustling the bag, as if there's something wrong with my hearing. 'Sweets. From Sweden.'

I pick one out and drop it into my mouth just as the vicar makes his appearance through a side door. He moves easily and elegantly through the obstacles in the aisle, and is soon ascending to the pulpit where he raises his hands aloft and looks out over the congregation.

'Everything has its time,' he announces in a friendly, but nevertheless deeply mournful southern accent. His gaze roams over the assembled company, row by row, until he ends with the closest relatives on the front row right beside the coffins. 'Even death.'

Once the prelude strikes up, people start to sing. *The day thou gavest, Lord, is ended, the darkness falls at thy behest.* Then the vicar delivers the eulogy. 'They devoted themselves completely to each other for many years, to love. And now they are with God, Our Lord, in Paradise, together.'

The vicar descends smoothly from the pulpit and moves across to the two coffins. The family sends two girls forward, both in their early teens and dressed in beautiful dark dresses with matching shoes. The vicar and these two girls begin to read out the messages written on the wreaths, as people step forward to light the candles in the wrought-iron candlesticks. The girls read in Swedish, and the vicar in Norwegian.

'Now Selma and Christine will sing a hymn for their grandparents.' The vicar folds his hands and moves to the side of the choir. Soon a melody rings out from a CD-player, and the two girls break into song.

The screaming child is no longer crying. The whole church is suddenly silent, with only a vague gurgling or sporadic cough from ageing lungs to be heard between the notes.

I can see the back of Merethe's head in front of me. She does not move: simply sits there with drooping shoulders.

The metal screws protrude on each side like an insect's antennae. I see Harvey at her side, his hands clasped on his lap with his face cast down on the floor.

He appears to be shivering. His shoulders are heaving uncontrollably, moving up and down, and from time to time he shakes his head, sometimes rocking back and forth with his hands on his knee. He is crying. Blubbering like a baby, and when the music is over and the two girls have left the choir, the whole church is filled with the insistent sound of his heart-rending sobs.

CHAPTER 56

'What do you think that was about?' I ask Johannes once the vicar has finished and people rise to their feet. The pall-bearers advance to take their positions beside the coffins. Only once they have hoisted them and are standing in line does it dawn on them that it will be impossible to convey the coffins up the crowded aisle. They stand looking at one another but no one seems to have any idea what to do.

A couple of care assistants from the residential centre see at once how serious the situation is and give a signal to the rest of the staff. They start to wheel the residents out, while others collect crutches and walking frames for those still seated. The two care assistants who brought Merethe have returned. They take hold of her chair, one on each side, and push her out with the others.

'He's been through a lot just lately,' Johannes says once the aisle is clear and the first of the pallbearers have come up beside us. Johannes gives a reverent nod while they slip past. 'After all, you were there yourself when his wife had that attack. It can't be easy for him.'

'It wasn't an attack.' I take a few short steps to one side, ready for our own turn to join the slow-moving mass of people.

'Well,' Johannes mutters, nodding once more as Harvey nears us. He is oblivious, staring vacantly into the middle

331

distance as he walks mechanically by, holding hands with his son.

Johannes gives his tobacco pouch a firm squeeze when we finally step into the river of sorrow. His eyes flicker over the people ahead of us in the burial procession as it turns to the left outside the church and sets off towards the graveyard.

Outside, the sun is shining. Its warm rays bathe the car park, church and the procession that has now gathered around the freshly dug hole in the ground at the end of the graveyard where the burial will take place.

We stand some distance away in the outer circle and can barely hear the vicar who has embarked on yet another bible reading. All at once a shadow stretches out towards me, and Harvey's grey face appears.

'Have you seen her?' He nods towards a rise beyond the last graves where the two care assistants stand flanking Merethe's wheelchair. She is sitting with her hands on her lap watching their son play beside a stream that runs alongside the church-yard fence. He is scooting up and down the bank, dragging a stick behind him with the tip immersed in the water, splash-ing in all directions. 'Her jaws are screwed tight, and they've stitched the muscles so that she has to be fed through a tube.'

'I'm sorry, Harvey,' I say, turning to face him. 'I—'

'No, Thorkild.' Harvey's eyes meet mine. 'I'm the one who should apologise. I was so upset when we last spoke, so frightened, but I should have seen the state you were in and insisted that you come with me to the hospital. Instead I started babbling about the lighthouse and ghosts.'

'Don't mention it.' I clear my throat. 'I'm the only one who can do something about that business. How is she?' I ask in an attempt to change the subject.

'Well, we're talking. At least, I talk, and she writes notes. In a week or two they'll let her home. The doctors say she will improve – things will sort themselves out in good time. I just hope she manages to make a recovery before the filming of her show starts. She's been looking forward to that for so long.'

The congregation has come to the end of the prayers, and the ritual lowering of the coffins has begun.

'You found the Dane,' Harvey says as the vicar performs the ceremonial scattering of earth before opening his bible and beginning to read aloud from it. 'When you were out there … in the sea,' he stammers. 'Tied to a woman's arm?'

I nod. The vicar closes his bible and moves on to the final blessing. 'Now at least nobody can claim that she doesn't exist.'

'Do they know who she is?'

I shake my head.

'Any news of Bjørkang and Arnt?'

'The police in Tromsø have requisitioned a mini-submarine and asked Kripos for assistance. Reinforcements are due this afternoon.'

'Do they still think you have something to do with it?'

'Very much so,' I answer. 'I'm going for another interview later this afternoon. There are just a few things I need to sort out in advance. See what's what, if you get my drift.' I take the apartment key from my jacket pocket and hand it to Harvey. 'Thanks for the loan.'

'No problem.' Harvey stuffs it into his trouser pocket. 'The relatives are clearing it out later today.'

'By the way, how did things go for you? Were you called for interview yesterday?'

Harvey shakes his head. 'I haven't been there. I can't manage yet. There's such a lot to attend to because of Merethe now. My God.' He bites his bottom lip and turns away: 'I've been so afraid for her.'

The burial is nearing its conclusion, and the assembled people are folding their hands in preparation for the final psalm. I place a tentative hand on Harvey's shoulder. 'May I speak to her?'

'Yes,' Harvey replies. Merethe has hidden her fingers in her jacket sleeves. She is watching the singing crowd with a fixed gaze. One of the care assistants produces a woollen blanket that he unfolds and places over her knees. 'Of course you can.'

We set off towards Merethe as the crowd begins to disperse, Harvey leading the way, with Johannes and me following behind. Above us the sky is clearer. The sea lies shining and still beyond the bay.

'Hi,' I say in an uncertain voice when we reach Merethe in the wheelchair. The care assistants withdraw a few steps, and one lights a cigarette as they turn to face a grove of trees behind the graveyard. 'I ... I ...'

Merethe holds a hand up, signalling me to stop. She opens one of the pockets of her suit jacket and takes out a notepad. While she writes, I steal a glance at the metal contraption again. Her lips are dry, with a gap between them. Her lower lip juts out beyond the upper one, making her resemble an assault victim with deadened nerve fibres and paralysed muscle tissue in her face.

Merethe hands me the notepad and looks at me. *It was not her* is written on the note. She nods her head very gently as I read.

It was so painful, it says. *As if my mouth was filled with more and more water with every breath I took. It felt as if I was drowning.*

'I tried to help you, but I didn't have the strength to hold ...'

Merethe shakes her head before writing something and hands me the notepad again.

It wasn't your fault. I was the one who let her in.

'I saw her too.'

Merethe nods as she writes. *She was speaking to you.*

'To me? What do you mean?'

Merethe writes again, and then tears off the note and places it on my open hand. 'I don't understand,' I whisper, staring at the letters that form two incomprehensible words, wondering whether she has suffered some kind of seizure that has led to her not knowing what she has written, nothing but random lines and letters laid out in a row.

'Looks like Russian.' Harvey looks down at the note. He is pale and his body is swaying slightly from side to side as if he is feeling faint. He has to take a step to one side, towards the wheelchair, to avoid falling over.

'Do you know what it means?' I ask.

Looking at us both, Merethe leans forward in the wheelchair. 'I'm freezing,' she wheezes through clenched teeth.

'What?'

Merethe grabs my hand and squeezes it hard. Her fingers are abnormally warm, so warm that I can feel the heat rise up my arm to my neck and face. Her chest heaves every time she gasps for breath, guiding my hand to the notepad and the two words written there. 'I'm ... freezing.'

'Did she say that? *I'm freezing*?'

335

Merethe nods and exhales. Letting go my hand, she writes something else on the pad and tears it off when she is finished, then passes it to me with a sorrowful look.

Goodbye, Thorkild.

'Goodbye,' I say, folding the note. I tuck it into my trouser pocket before we all exchange polite phrases and go our separate ways.

Johannes and I head off towards the car together with stragglers from the burial. The sun has now broken completely through the cloud cover. In front of us the last car rolls out of the car park and on down to the main road.

'Johannes,' I say as we approach the hire car. 'Arnt told me about a steady rise in narcotics and prostitution in Tromsø on the first day I arrived here.'

'Yes,' Johannes answers. 'You have only to read the papers, you know, then you can see where things are going.'

'Do you know where these women come from?'

'Yes,' he says, laughing bitterly. 'The same place as the booze, cigarettes, tobacco and the rest of that shit. From Russia.'

'Yes, true enough,' I concede, and get into the car and start the engine.

CHAPTER 57

'What are you hoping to find out there?' Johannes asks when we park the hire vehicle beside the boathouses.

'Answers,' I mumble cryptically, surveying the still bright sea between the bay and the lighthouse.

'To what?'

'Who the woman without a face is. Who murdered Rasmus. Where Bjørkang and Arnt are, and maybe confirmation of whether I'm crazy or not,' I tell him before I open the door and step out.

'Is that all?' Johannes chuckles to himself.

The salt sea wind brushes over my face and lightly pricks my skin. I stride quickly back to the car, open the boot and change my clothes again.

It takes just a few minutes to cross. Johannes steers the boat in towards the quay stump beside the lighthouse and goes alongside. Grabbing a coil of rope, I clamber up and tie the boat to a rusty reinforcement iron sticking out of the bedrock.

Johannes looks at me on my return to haul the old fisherman ashore. 'What now?'

'The main building. The disco is in the basement. After the séance with Merethe I came out here.'

He comes to a halt at the front door of the renovated keeper's residence. The crime-scene tape I tore off last time I was here is lying crumpled beside the wall.

'Séance? What séance?'

'There was someone in the room with us, inside her. She was talking, screaming.'

'What do you mean?' Johannes stammers, as he stands frozen in the doorway while I step into the foyer. 'A … ghost?' he calls out after me.

I come to a standstill, looking at him, aware of the rustling plastic and smell of old mildew and foul air filling my senses. 'You must have known that Merethe is clairvoyant?'

Johannes spits and sniffs loudly. 'I don't like that sort of stuff,' he says, treading reluctantly into the foyer. 'That kind of thing gives me the creeps. Some things are not meant for us to fool around with.'

We pass the plastic-covered walls of the foyer, heading to the staircase. 'What did she say?' Johannes asks when we get there.

'Sorry?'

'You said she was talking.'

I take out the note Merethe gave me and hand it to Johannes. '*Mne xólodno*,' I say.

'What?' He looks quizzically at me. 'What did you say?'

'*Mne xólodno*,' I repeat. 'It means *I'm freezing* in Russian.'

Johannes's face blanches. 'It's impossible,' he whispers, staring vacantly into space.

'What is it?'

'No, no,' he mumbles, almost in a daze, as he grabs at the banister for support.

'What is it, Johannes?' I put a hand on his shoulder.

'I ...' he starts to explain and then his gaze returns to me. His eyes are wide open and his mouth is trembling as he speaks. 'It's just that I think I've heard this before.'

'Where?'

Johannes takes a deep breath before straightening up again. He digs out his tobacco from his jacket. The first cigarette paper tears as he tries to extract it. 'There was a boat,' he explains, screwing up the paper in his hand. 'A Russian boat that went down here a few weeks ago – I don't know whether you've heard about it?'

'Bjørkang mentioned it. What about it?'

'It had engine trouble en route to Tromsø. There was a terrible storm, and the boat went down. All the crew managed to get ashore and they were taken to Tromsø that same night.'

Johannes produces another paper and fills it with a strip of tobacco. He rolls the paper slowly and methodically, the cylinder shape emerges, and he lifts his creation to his lips to moisten the adhesive. The roll-up sticks to the tip of his tongue and it tears. 'I heard them on the walkie-talkie,' he says in a musing voice. 'First there was just a lot of shouting and cursing in Russian, and then they changed to speaking in English – a few short sentences when they made contact. Then it all went quiet.'

'Do you know who they were talking to?'

'No. Too much interference on the line, and just after that it went completely quiet.' His tone turns sombre, as if what he wants to say is scaring him. 'Until later that same night. That was when I heard it.'

'Heard what?'

He takes out a fresh cigarette paper while he talks, and adds a strip of tobacco. 'I had just been out to lay a heavy stone on the cellar hatch so that the wind wouldn't blow it open and rip the whole thing off. When I came in again, I heard a crackling noise from the living room. I took off my boots and went inside. The green light on the walkie-talkie was flashing, as if somebody was sending a message, or had at least pressed the send button, without saying anything.'

'Someone from that boat?'

Johannes observes me with eyes that are narrow and black. His lips curl down, as if trying to turn away from the rest of his face. 'I stood there in front of the walkie-talkie, waiting, to see if the person sending a message would try again. I was wondering who it might be.' He looks at me with that strange expression on his face, a mixture of fear and amazement. 'I mean, it sometimes happens that some-one sitting on one of those godforsaken rocks out here gets a bit down in the dumps in the wee small hours, and you can expect to hear anything at all at that time of night. But something was different; I could feel it every time the green light came on. It wasn't until I tried to answer, that ... that—'

'What?'

'I heard a voice. Up until now I've told myself that I'd just put a drop too much of the hard stuff in my coffee that night, or else that it was the feeling you get sometimes when you're sitting there on your own in the house while storms and gales blast and tear at the walls; that what I heard was just atmospherics, or wind or whatever. But now—'

I realise that I have gone cold inside. As if my brain already knows what Johannes is going to tell me, and is warning my body that I will soon be feeling even colder. 'What?'

'It was a woman's voice.' Johannes's face contorts into a grimace of pain. 'She was whispering, so softly that if it hadn't been for the green light it could just as easily have been the wind sighing. Two words in a foreign language, and then the light was switched off, and I didn't hear anything more from her.'

'My God,' I exclaim, taking a deep breath. 'What are you telling me?'

Chapter 58

'Are you sure you want us to go down there?' Johannes has finally managed to light his roll-up. He looks anxiously at me through the tobacco smoke.

'Yes. I have to know,' I answer, leaning down over the banisters in an effort to see the foot of the stairs, where the door was left open a crack last time I was here. The line between fantasy and reality ran down there. 'There's no other way.'

We linger there for a while, each lost in his thoughts, as we both stare into the darkness below. In the end Johannes stubs out his cigarette and tucks the butt away inside his matchbox. We swap glances one last time and then descend.

The metal door at the foot of the staircase stands open. Inside is darkness and utter silence. I take out my mobile phone and switch on the flashlight function as we enter the corridor with the bird-display cases. The footprints from my last visit are covered in a fine layer of dust, and I follow them along to the wall with the stuffed birds on show. In the dim light they resemble stage props in an American horror movie from the seventies, their dead glass eyes squinting out.

'I've never been here before.' Johannes shudders as he approaches and catches sight of the exhibits.

'It's not exactly your type of place,' I mumble, continuing past the glass cases and across to the disco entrance, where I pause in the doorway. Total silence envelops us. The music, disco ball, strobe light and smoke machine are all switched off. The odd particle of dust continues to swirl around beneath the ceiling. The strong pastel colours on the concrete walls are patched with the same grey damp: a desolate moonscape.

'Come on,' I say, pressing further on into the disco. 'She was sitting over there.'

We pass the DJ booth and stop in the centre of the dance floor, where I can barely make out the green glass with the male figure pointing in the direction of the emergency exit.

I feel my heart sink as I come closer. 'Empty,' I say, stopping in front of the booth where the table is covered in dust and the jam jars with tea lights are gone. The air is rank and oppressive. 'She's not here.'

'Thank God,' Johannes gasps in relief. 'For a moment there I was really feeling my heart start to pound.' He fumbles to find the cigarette stub in his matchbox, but changes his mind and puts it straight back again. 'What a stink.' His lips curl, and he raises the hand with the matchbox to his face, as if it were a pomander.

The light from the mobile torch encounters an indentation in the flattened dust on the sofa, where I sat last time I was here. In the opposite booth there is a patch of dark dried matter on the seat where the woman with no face had sat. 'But she was,' I murmur, squatting. I can make out a puddle of sticky, rancid paste that looks as if it has run out from under the table. Covering my nose with my free hand, I breathe through my mouth as I stoop closer. 'She was here that night.'

'What is it?' Johannes peers impatiently around, holding his nose. 'Have you found something?'

I point the flashlight at him before turning the beam back on the floor. 'Footprints,' I answer.

In front of me I can see my own wavering prints in an arc from the booth to the dance floor before heading in a straight line, ending at the emergency exit. I can also see a thicker line that starts at the booth where I am sitting and partly passes over mine before these also stop at the emergency exit.

'It looks as though someone has come here and taken her out the same way that I left,' I say. I swing round and play the light over the floor from the booth, and in the opposite direction, further in, behind the DJ booth. 'Not long after me.'

'How do you know that?' Johannes's breathing is growing more and more laboured and his face looks greyer the longer we stand down here. He shuffles his feet and rattles the matchbox impatiently on his trouser leg as he speaks: 'You can hardly see anything at all in here. There's just this foul stench—'

'The dust,' I answer, pointing at the floor ahead of me where two sets of footprints are visible in the soft layer of dust. 'My own prints and the other set have exactly the same amount of dust on them.'

'Did you see anyone while you were here?'

'No, but I doubt whether I'd have noticed a drunken elephant jiving on the dance floor, the state I was in.'

Johannes tries to laugh, but his smile falters, fades. Instead he presses the matchbox containing the cigarette stub against his face once more and follows me. I trace the footsteps with the mobile light held out in front of me, past the rows of empty, dust-covered seating covered in rotted upholstery fabric, the DJ booth, and on in towards two billiard tables in one of the corners.

The footsteps end behind one of the billiard tables. The dust is trampled in a pretty large, irregular circle. The baize covers are only just visible and billiard balls lie spread out on the tables. They look like small cabins in a miniature landscape of the type found in a natural history museum dedicated to lost civilisations.

I direct my gaze up and out across the room. On the opposite side of the disco, I make out the emergency exit with the green man and the booth where I had sat last time I was here. On the other side I can barely discern the light from the entrance area. In the middle, between the two, is the dark wall of the DJ booth.

'An ideal lookout point, don't you think?' I suggest when Johannes crosses to stand immediately facing me.

'For whom?'

'A culprit,' I mutter, shining the light on a door that is open a crack, between two billiard-cue stands at the far end of the room. 'Who enjoys playing with me,' I add, going over to the door and opening it wide. Inside, a dark, narrow staircase leads up to the ground floor.

'Come on,' I say. 'I've seen enough.'

———

'Did you find what you were looking for?' he asks once we can finally close the basement door and leave the rancid darkness behind us. We have ended up in the restaurant kitchen on the ground floor. The room consists of a food store and three freezers on one wall, one of them humming faintly. The room has been recently scrubbed down and smells strongly of detergent and cleaning agents.

'Absolutely,' I answer, turning to Johannes with a smile.

'And that is?'

'I think the dead woman was Russian,' I explain as we continue to walk through the restaurant, where furniture is stacked and covered in plastic and white tablecloths in one of the corners. 'Maybe she was a sex worker on her way to the city to find work – that would at least explain why no one has reported her missing. I think Rasmus found her while he was out diving, and he was murdered because of that.'

'Good grief,' Johannes grunts, rubbing his hands when we are finally out in the foyer again. 'Where will all this end?'

'And last, but not least,' I round off, mostly to myself, with a hint of a smile on my lips. 'The rumours of my own unreliable state of mind have been greatly exaggerated ...'

CHAPTER 59

'Well.' Johannes has returned the matchbox to his pocket and now claps his wrinkled hands, as if to congratulate himself on being outside again. At last, having left the main building, we have stopped in the yard between the keeper's residence and the boathouse. 'Then surely it's time to go back ashore again?' His eyes are squinting at me. 'Think of dinner.'

'This trawler,' I say, tucking my mobile into my jacket pocket. I perform a few short facial gymnastics to chase away the tingling under my skin. 'Do you know where it sank?'

'Of course – I plotted the coordinates in the GPS to see where they were when the SOS came in. I have a portable, you see.'

'Is it far from here?'

'No. Near the northern tip of the island.'

'Deep?'

'Not specially.'

I shiver as we both let our eyes wander across the surface of the sea, smooth as glass, beyond the little island where the lighthouse is located. 'I think I need your help,' I tell him, after a few moments of silence.

'I thought you might.' Johannes screws up his eyes as the reflection of the sun on the sea strikes his face. 'You're going to need equipment,' he eventually comments.

'We've got everything we need here.' I turn to face the boathouse. 'In there.'

'Can you dive?' Johannes takes the half-smoked stub from the matchbox and lights it once we reach the boathouse.

'Yes.' I pull the door open and make for the emergency generator exhaust system and the pipe where the plastic sheeting – that I'd used to wrap the woman's body after I hauled her out of the sea – is lying. I continue on towards the cases of diving equipment and take out an inner and outer suit, shoes, gloves, flippers, bottles, brackets, carbine snap hooks, a diver's lamp with pistol grip, and another lamp that attaches to the actual mask, as well as a knife.

Apart from the obligatory mine-clearance diving courses in the military, I had also – unwillingly – accompanied Gunnar Ore on a couple of team-building trips to his family home at Nesodden while I was still serving in Internal Affairs. We all had to take part in either wreck diving or underwater fishing for the duration of the trip: even our legal team were forced to plumb the depths.

'What makes you so interested in this wreck?' he asks once I am ready. He sucks down the last traces of smoke and fritters the embers between his fingers.

'I think the woman I found in the sea out at the lighthouse was on board the trawler that sank on its way to Tromsø a month ago. I think that Rasmus may also have heard them on his walkie-talkie when they sank, and decided to dive down to the wreck the weekend he disappeared to take a look. So he found the woman. That is why no one can know about her, and why the GPS equipment was removed from Rasmus's boat and from the ambulance vessel used by Bjørkang and Arnt. Because they've both been out there at the wreck. They didn't all get out of that boat before it sank – she was left behind.'

'Rasmus was killed because he found her on board the wreck?'

I nod.

Johannes shudders. 'Does that mean that she was the one I heard on my walkie-talkie that night?'

I press fingers into my cheek while the other hand instinctively grasps my jacket pocket where the insect eggs are usually kept. Now my pocket is empty.

'B-but,' Johannes stutters when I don't answer. 'That was a long time afterwards. The boat must have already been sunk and …'

Our eyes meet, and he doesn't complete the sentence. We stand there in silence for all of a minute until finally we pack up in the boathouse and carry the bags down to the boat.

'I still don't understand what you're expecting to find in this wreck,' Johannes says once we have set off back to the boat.

'First of all we have to find the vessel,' I tell him as a chill gust runs through my body and combines with the gnawing in my belly that had started as soon as I discarded the empty blister pack of Paralgin Forte on my way into the church. I pull my jacket collar more snugly around my neck and swallow, in the hope that the cold air will pump my withdrawal symptoms further down into my gut. 'But once that's accomplished,' I say in an undertone, 'there's one thing I'm absolutely sure we'll find inside it.'

'What's that?' Johannes asks as we clamber on board and take our places.

I look at him. 'A dead policeman.'

Chapter 60

We hug the shore of the main island, heading northward past bare hillsides, small islands and rocky skerries, steep mountains and cliffs, as well as the occasional sand- or pebble-fringed bay hidden in amongst them. The breeze ruffles my hair and tweaks at my face.

'What is that?' I point towards serried ranks of floats that have appeared outside a narrow cove just ahead of us. The floats are suspended side by side in six lines with approximately two metres' distance between them. The entire area is around two to three hundred metres long and runs at an angle across the bay.

Johannes drops his speed so that we are moving more slowly through the water. 'The farm,' he says. We see a barge and a substantial blue aluminium catamaran moored close to the shore. Farther in, there is a modern quay with a fairly large boathouse. The cove is tucked between two squat hills flanking a white-painted timber house and barn. The place is dark, as if the sun did not reach the farm because of the closeness of the hills. 'Harvey's mussel farm. You can see Merethe's childhood home on the shore.'

'I don't see a factory,' I remark as we pass over the crystal-clear shallows. 'Where does he process the mussels?'

'On the catamaran,' Johannes answers. 'He uses the boat for setting out the spawn, harvesting, cleaning and packing, before he transports them to the warehouse in Tromsø, and from there they're dispatched on large pallets out into the big wide world. That's actually all you need nowadays.'

'It looks solid.' I nod in the direction of the blue, purpose-built boat with a massive, heavy crane at the stern, as we pass the quay and proceed out of the cove on the other side.

'Boats with side moorings, you know. That one there can harvest mussels in all kinds of weather.'

In front of us a number of smaller islands and reefs come into view, where a group of birds with bluish plumage, giving an almost metallic sheen, and white markings on their heads, are perched, drying their wings in the good weather. Some have their heads bowed, while others crane their necks at the boat as we approach.

'I don't like those birds,' Johannes says, shivering as we close in on the largest islet where a slanted iron post holds a beacon on a submerged reef, just offshore. The yellowy tufts of forests of seaweed stir on the surface of the sea all around the little islands and nearby reefs. A few of the birds flap their wings in the water before flying off when they appear from behind a mound of bladderwrack.

'What? Why?'

'The cormorant warns of death,' Johannes mumbles, accelerating again as soon as we have passed the last of the reefs.

'Oh, bloody brilliant,' I grunt in response, sinking my head down between my shoulders as we pick up speed and the boat begins to thud against the waves again.

'Hold tight,' Johannes shouts. 'It's not far now.'

I grab hold of my seat, using both hands. Farther ahead, I see how the entire island tapers and curves in on itself. In the distance there is nothing but sea.

After a while we have passed the tip of the island, and Johannes reduces his speed in preparation for steering round in a figure of eight, keeping his eye on the GPS and echo sounder between his swerving manoeuvres. 'Here,' he announces at last, switching the engine off completely, so that the boat settles, rocking lazily in the slight swell.

'Have we arrived?' I get to my feet and peer over the gunwale. The sea is glittering in the dappled afternoon light. We are about a kilometre from shore. The sky has assumed a blood-red, almost purple, shimmer.

'This is the location of the GPS coordinates they gave on the walkie-talkie. The echo sounder indicates that the seabed tilts down slightly and gets deeper farther out. We don't have much time,' he says, scanning the sky and looking back at me again. 'It'll soon be dark.'

Nodding, I take out the diving suit and put it on as fast as I can before the autumn chill penetrates my bare skin.

Johannes glances at the time before peering over the gunwale, down into the cold water. 'I hope you're wrong. That there isn't even a boat down there.'

'I hope I'm wrong too,' I answer, while checking the pressure in the air tanks. 'Believe you me.'

I put on the diving vest, take hold of the mouthpiece and check the regulator, and then add the diving mask, switching on the light. The truth is that I hate diving. Hate it with a vengeance. But this is one of those things you simply can't avoid, such as rectal examinations when you have a sore backside, and having your thoughts ransacked by a shrink when you've attempted to take your own life.

Inserting the mouthpiece, I pick up the lamp with the pistol grip and check that it's working before sitting on the edge of the boat with my back to the waves.

'We can always turn back,' I hear Johannes say as I lean backwards and tumble over the gunwale. My body breaks the surface of the water almost immediately.

I think it's too late is what runs through my mind the moment the cold surf hits the skin between the hood, mask and mouthpiece. *Far too late for that now.*

CHAPTER 61

The cold makes my face contract. Chilled and floundering, I kick out to get my circulation moving. I have never dived in late autumn before and struggle to adjust the respiration and force enough heat into my body.

After some time spent just below the surface, kicking hard and thrashing about, I eventually muster rhythm and calm enough to get moving. Above me I can see Johannes leaning over the gunwale. I give him a thumbs-up before turning round and starting downwards. Next time I look up, he is merely a dark shadow far above.

The water is clear, but rapidly grows so dark that I have to use the hand-held flashlight to see more clearly. I can already make out the slope of the seabed below me. The depth gauge shows seven metres. The effort of controlling my breathing and the unnatural movement of my thighs makes me think of Frei and me at the dancing class in the Arts Centre the day I followed her. It occurs to me that this is the first time I have thought of her since that evening out at the lighthouse.

'You there. Come on!' commands the instructor, strutting proudly in front of us as we sit in a semicircle around her. She is wearing high heels, a short black leather skirt, and her dark curls are tied in a ponytail. She reaches a hand to me, as if to summon a scruffy dog.

Reluctantly, I stand up and approach her, leaving Frei seated on the floor.

'OK, *señor*, what's your name?'

'Thorkild.'

'*Señor* Thorkild. Have you danced before?'

'A little.'

Now she turns to face the group. 'OK then, *amigos*. Pair up, and *Señor* Thorkild and I will walk through the basic steps with a simple shoulder rotation, and also demonstrate some arm movements and back-to-back.'

'Imelda?' Frei comes across to us in the centre of the circle, where I stand held in Imelda's iron grip. 'Couldn't you take Robert instead? And then I can have this man here?'

Imelda lets go of me and looks at Frei. 'Robert? *Si*,' she laughs. '*Claro, hermana.*' Then her face stiffens again, and she gestures to Alvin, who makes his way across to the stereo system.

'Which song, *Señora* Imelda?'

'"Dos Gardenias", *Señor* Alvin.' Immediately afterwards, a silky soft male voice singing in Spanish rings out as Robert glides elegantly past me into Imelda's arms.

'What are you doing here, Thorkild?' Frei whispers as she leads me in among the other couples.

'I don't know,' I answer. Frei puts her right hand in mine, while her left grips my upper arm.

'Has it anything to do with Uncle Arne or that other case of yours?'

'No.'

'I see ...' She hesitates. 'You came down here just to see me dance?'

'I believe so,' I murmur as I struggle to keep pace with the rotation before Imelda's voice sounds again above the music.

'Come on now, all of you. Hips, Thorkild. Hips. This is a bolero. Hot, torrid, sensual, *calma, calma*.'

'There's a word for that sort of thing,' Frei goes on to say when Imelda has finished her instructions. 'For people like you.'

'I know. Pathetic,' I sigh. 'You ...' I'm about to pull away when Frei's fingers slide farther into mine, and I can feel her breath on the hollow of my neck, and her eyes burning on my skin. Her face is only a few centimetres away from mine. All I had to do was to lean in and kiss her.

'Or dedicated?' Frei continues. 'Inquisitive?'

'Too old.'

'For what?'

'For this.' I make a fresh attempt to free myself, but Frei will not let go.

'Don't go,' she says, holding tight. 'Not while we're in the middle of the dance. You can leave when we've finished.'

'OK,' I answer, and then draw another breath when Imelda signals that a new rotation is coming. 'We'll dance.'

This moment always stops just here. It has no sequel, even though I know that we parted shortly afterwards. She and Robert went one way and I headed back to the hotel room and my interview documents, reports and whole writing pads of notes about things that have no meaning at all. Everything stops here, even me, even though right now I am drifting under the surface, searching for a boat on the ocean bed.

'You bloody fool!' I gasp, and next second my mouth fills with salt water. I open my eyes and see that, in my bout of self-reproach, I have managed to spit out the mouthpiece connected to the diving bottle.

I can still smell the scent of her hair and feel our dance moves as I cast about for the mouthpiece and force the water from my mouth, all the time struggling to hold my breath.

After a couple of rounds of fumbling, I catch hold of the mouthpiece at last and insert it again. I hang with my arms outspread like an angel as I concentrate on regulating my breathing once more. Just after that I feel my feet touch bottom.

Kicking out to avoid sinking into the clay, I check the dive computer and pressure indicator. I've reached a depth of eleven metres now, flanked only by a couple of dead fish partially buried in the soft clay on the seabed with one or two crustaceans nibbling at them.

The bed slopes down at an incline of roughly thirty degrees. In between, I can make out the occasional stone or piece of scrap metal protruding from the mud, providing a home to small colonies of sea algae, starfish, sea urchins, mussels and sea snails in the sediment.

I am still freezing, but have hit upon some sort of rhythm as I drift slowly down into the darkness. The seabed must be the loneliest place on earth: barren, murky and alien. I've never felt more lost and alone than right now, as I drag myself across mostly featureless mud towards a dark wall looming in front of me, jutting out from the seabed. A grey cloud of agitated sediment blankets the wall, and it is not until I'm inside this cloud that I realise that the wall, far from being a wall, is in fact the wheelhouse of a boat.

The wheelhouse is lying upside down, resting on boulders or bedrock under the mud. From a gap in the top where the wheelhouse has been severed from the rest of the boat, I can see thick pipes, fragments of cables and other wreckage sticking out.

I swim all the way over and up along one of the walls until I'm poised just above the opening. I can see that it would be extremely tricky, not to say hazardous, to attempt to wriggle through the cluster of pipes, cables and wreckage inside.

I move on past a flight of rusty stairs. Already, mussels, algae and other forms of sea life have begun to stake their claim to the white metal walls containing the row of windows on the wheelhouse's bow end. Above the windows, I can make out big Russian characters that must display the name of the trawler.

Brushing scum from one of the windows, I peer inside. The first objects I catch sight of are the chairs projecting from the ceiling that had previously been the wheelhouse floor. A coffee mug floats in an air pocket just above the instrument panel. Further back, I see the outlines of an international sign displaying the on-board locations of the lifeboats and how to don lifejackets.

I hang with my diving mask pressed firmly against the glass as my eyes slide over the loose items swirling inside. Suddenly I glimpse a flat-headed fish with a prickly dark-green back and bulging eyes emerge from a hole in the instrument panel. It propels itself slowly up towards a paper cup and tugs at it before swimming around and trying again from the other side. Our eyes meet for a fraction of a second, before the ghastly creature turns and darts between the joints on the wall panel.

Backing away, I check the pressure indicator before scanning the area for a way out of the fog churning around the wreck so that I can read the measurement. I catch sight of a thick cable, seemingly attached to the wheelhouse base, that plunges further into the darkness.

I swim across and put my hand on the steel wire. The threads vibrate with the tension caused by whatever is pulling at the other end. I check the pressure indicator and safety valves one last time before using both hands to grab the wire and start to haul myself down into the gloom towards whatever awaits me at the other end.

CHAPTER 62

Gradually as I descend, more dead, rotting fish and increasing amounts of wreckage appear, scattered across the seabed. I can also see something that looks like one of the stern beam-trawl nets protruding from the mud. The boat must have rolled down into the depths, scattering tackle, plastic drums and flattened freezer boxes on the way.

At a depth of twenty-seven metres, I stop to check the depth gauge and pressure indicator one more time. In front of me, another wall of dark steel emerges from the sea floor. The trawler is lying with its keel upturned and almost straight. It seems as if the cable is the only thing preventing it from rolling farther down into the depths. The whole area is enveloped in the same grey fog as the wheelhouse farther up.

Cautiously, I swim closer and stop immediately in front of the keel where an enormous trawler net is spread out. The boat must be about forty metres in length. The keel is covered in small wart-like protuberances, a scattering of grey-green tassels of seaweed, and the occasional cold-water coral. I can also see sporadic colonies of dead man's fingers with yellowish-white and cream-coloured nodules waving at me from the underside of the wreck.

The only sound is the bubbles emitted by the valves at the side of my head each time I breathe. The murk surrounding the wreck has captured both the trawler and myself in its own world down here on the seabed – a cloudy, grey-ish-brown, algae-sprinkled fog that pushes and pulls on you from all sides, testing you, as if to see whether you and your armour will really hold out and can bear the strain.

The entire ship has lurched dangerously forward and might roll farther still at any moment. It is open between the bow and the sea floor, so that I can swim round and come out on the deck side. All the same, I choose to swim up and across the hull for fear that something will give way and the whole metal monster will shift and crush me under its weight.

I start to heave myself up from the keel until I reach the jutting upturned windlass on the port side. The anchor is gone and only the chain remains, stretching out from the spool and disappearing down into the depths together with the front mast.

A swarm of algae drifts past the wreck just beyond the patch of fog, moving up towards the surface, as if flee-ing from the darkness down here. *It would have been so easy*, I think as I continue past the anchor winch in the bow, progressing astern along the side of the vessel. So why am I suddenly so afraid? Why is my body suffused with an intense fear of not returning to the surface now, when I am finally aware that all I need to do is to yank out the mouth-piece and inhale? It suddenly registers that this is what it looks like, the end of the line that I've been imagining for so long. As cold, colourless and desolate as this trawler at the bottom of the sea.

I grab a piece of metal by the railing as I close my eyes and try to breathe calmly. In the end I regain control over my respiration and swim onwards. Ahead of me I see the front of a shattered lifeboat sticking out from underneath the railing on the trawler's starboard side. Part of the cold-storage plant is suspended from a large hole where the metal has cracked, forming a broad split across the deck. Onwards, Thorkild. Only onwards.

The seabed is almost white with dead fish, some still contained in freezer boxes or packed in plastic blocks. It seems as if the pile is alive, changing colour from chalk-white to deep-green, ruby-red and shades of yellow. Glittering streaks of silver shoot between the pulsating lower layer when small shoals of fish fry scurry past the heaps of crabs, starfish and sea urchins that have arrived to join the feast.

I continue backwards, past the trawler ramps, rows of filleting stations, nets and loading hatches towards the stern, where the wheelhouse once stood. In the murk, huge frayed metal edges jut out from the deck as well as pipes, cables and other wreckage.

At the stern I remain poised above the opening and the two gaping holes beneath me, each with a rusty metal staircase leading into the bilges of the ship. One goes down to the cabins and galley and the other to the engine room and technical areas.

The winch – the wire I followed down from the wheelhouse was part of this – is hanging in front of me. The mounting is broken, and the winch arm is bent down towards the deck, while the spool itself is still attached to the supports in front of the wheelhouse remnants. The bolts on one side are torn loose and there are two black holes where the brackets would once have been located. The cable winds down from the spool across the deck on the starboard side, beneath the boat and out at the back. As soon as the last two bolts tear off or manage to twist their way out of the metal that encircles them, the entire trawler will continue its journey down into the darkness.

I am aware there is a real risk that the wire will give way or shear off, and that the wreck will then tilt over on one side to lie with its keel up, or will simply slide down the slope, and that I risk being trapped inside if I dare to approach any closer. None of these propositions seems especially attractive as I hover above the two staircases, but at the same time something draws me into the gloom.

I try to look beyond the cloud of dust and sediment up towards the surface of the sea, but all I perceive are faint shadows of grey and blue. 'Bloody fool,' I mutter to myself in faultless submarinish, without detaching my mouthpiece, while I lower myself carefully towards the wreck below. When I am close enough, I begin to swim across to the staircase leading down to the engine room, where I stretch out my arm and reach for a length of rail to hold on to.

I direct the flashlight beam into the darkness and see that it is impossible to advance any farther than the end of the staircase. Heavy pieces of metal, engine parts, cylinders and other debris have collected just inside the cavity, blocking access.

I continue across to the other cavity and shine the light inside: an orange raincoat and a boot are floating around in the cramped stairwell that slopes upwards into the bilges of the ship. The way to the cabins is open.

Tentatively, I start to haul myself up the steps towards the aperture, concentrating hard on breathing well and keeping my cool. As soon as I have negotiated the stairs and thrust my head in through the door, I am faced with clothing and personal belongings floating around in the corridor ahead of me.

I feel remarkably peaceful as I move from one door to the next, looking at bed linen, books and framed photographs of women and children. Everything is swirling around in the bunk alcoves, peopled by the occasional tiny fish that has strayed in here. As if I am no longer so alone.

In one of the cabins, a pink bag drifts around under what had earlier been the floor, as well as a hooded sweater and a pair of trainers. There are no little family photos or pictures of naked women pinned on the door of this cabin. Toiletries, a pocket mirror and a few Russian fashion magazines, and a book with a picture of a man and woman locked in passionate embrace on the cover, are what glide between my legs in here. This cabin does not belong to a man.

I check the manometer to see how much air I have used. I realise my body temperature has dropped, and that working my way through these narrow passages is feeling more cumbersome now.

I pivot round and swim back out of the cabins, and as soon as I am out again, I swim down to the base and centre of the boat deck where the hatches for the cargo tanks and the fresh-water tank are situated. One of the tanks is open.

I haul myself across the deck, over to the open hatch, and peer inside. There are more dead fish, huge stacks of boxes packed with dead fish, and sheets of transparent plastic hanging from massive reels fastened to the wall. Tiny bubbles rise from the heaps of rotten fish still piled up inside the tank.

The next hatch is closed on the outside. I open the lock and hoist myself up so that I can press my whole body against the handle. In the end I manage to turn it far enough down so that the hatch springs open. I kick out to rise above the hole in order to avoid being buried in this tank, which is also full of fish.

I remain there for some time before thrusting the hand with the flashlight into the opening and taking a peek inside. As soon as I come close enough, I have the uncomfortable sensation that I am about to intrude on someone else's privacy.

The tank seems to be open. Even though I have both the hand-held flashlight and a smaller LED light attached to my diving mask, it is as if the darkness in here can smother the light in some special way. It feels as though I am peering into a reservoir with walls of oil or tar.

Cautiously, I descend into the tank. As soon as I'm inside, I hear a series of knocks resound throughout the wreck. A cloud of air bubbles drifts past, enveloping me so that for a moment I cannot see anything at all.

My body aches to get out of here, and every stroke and kick inside this gloomy fish tank fills me with panic and anxiety. As soon as the bubbles have passed, I continue my journey up through the tank. Below me, the light from the opening diminishes, as fine rust flakes loosened from the bottom of the wreck rain down through the water.

I stop halfway through the tank and hang there suspended as I focus on controlling my breathing, which is becoming erratic again. As I hover, with my head tilted up slightly as I watch the shower of rust, I glimpse something above my head. I straighten up and check the oxygen gauge before adjusting the LED light on my diving mask. Then I lunge towards the lifeless body above me.

The water sparkles faintly with myriad bright rainbow colours where the torchlight strikes it, as if contaminated with petrol, diesel or oil. The sparse layer of colour clings to the corpse rocking lifelessly in the rhythm of the water's motion.

It appears to be floating in thin air against the rusty backdrop flaking off and raining down on us: a black underwater angel with head, arms and legs reaching down towards me as in a greeting between two disparate states.

But everything is completely different now that we meet again.

CHAPTER 63

I did not think that he would be the one I found down here. Arnt Eriksen, the local police sergeant, still carries a bottle of air on his back as he drifts around the tank in full diving gear. His face is wrinkled and bloated, and his bulging eyes stare vacantly out. His mouth gapes open and his lower jaw trembles slightly when I move closer. A tiny brown sea creature is trying to hide in his trim moustache, but eventually gives up and slithers inside the dead sergeant's nostril instead.

I kick out yet again to reach the same level as the dead body, and next minute I break the surface to find myself inside an air pocket between the hull and the water in the tank.

The flashlight beam and the fine layer of petrol or oil on the surface make the confined space shine, almost as if we are in a lagoon surrounded by sun and white beaches. Not until I look up at the rusty keel a metre and a half above me is the illusion shattered.

I dive and carefully grasp the sergeant's shoulder, letting the torchlight slide over his spine and the back of his head, but there are no signs of external injuries. Then I embark on the difficult job of turning the body round to examine the underside. This proves almost impossible: the body simply reverts to the same position each time I try, and in the end I decide to remove the diving bottles before I move him round.

I twist the body into place before hauling the harness off on one side – an exhausting task. Not only is the body heavy and tricky to manoeuvre, it is also extremely arduous to keep working my legs to stay afloat at this level and at the same time struggle with the heavy body, on top of the exertion involved in avoiding gasping for breath or relinquishing the mouthpiece as I do so.

Following a lengthy pause, when I simply lie quietly beside the corpse, I turn the body, and after a great deal of back and forth and complicated manoeuvring, finally succeed in removing the other harness. Another period of rest ensues before I manage to haul the body with me up to the surface.

Somewhere behind me I hear hollow clanking from the air bottles bobbing on the surface as they jostle against one of the tank walls. As soon as I ascend to the glittering air pocket beneath the rusty dome, I start to drag the sergeant over on to his back to examine him in more detail.

My thighs ache and my arms and shoulders feel numb. It is awkward to get a good grip on the slippery diving suit and simultaneously keep the body above water. As I struggle to rotate it all the way round, I let go and the corpse sinks rapidly through the film of oil or petrol under the water again.

I grab the flashlight and am about to dive down to bring him up again when I glimpse something farther off. I hover here for a short time with the light directed at the silhouette. It dawns on me that I have made a mistake as soon as I swim towards the other body ahead me. A big mistake.

Police Chief Bendiks Johann Bjørkang is wearing a raincoat with a hi-vis vest and boots. The jacket is open, and underneath he has a blue shirt and tie. His eyes and mouth are closed, as if he is lying fast asleep here in this cocoon of water.

A cluster of starfish has collected around a point on the top of his head, just the same as the one evident on Rasmus Moritzen's body during the post-mortem. I carefully brush some of them away with the flashlight. The spiny creatures jostle their arms and suckers against one another as they drop to the bottom.

The grey indentation at the top of the skull is surrounded by hair and torn skin where someone has battered the skull with a hard instrument. There is little doubt that Bjørkang was dead when he landed in the water. I can see that his hands are joined with cable ties, the same kind of white plastic strips as on Rasmus.

One did not murder the other, or engineer their joint disappearance. They had never run around manipulating evidence and trying to put the blame on me or anyone else. Warily, I turn the local police chief's body round again so that he is lying flat and face down.

Suddenly I spot a movement all the way down at the entrance to the tank. At first it looks like the shadow of a big fish sliding in through the hole, before it stops for a second, turning round and then shining a bright beam of light up through the water to where I am suspended near the surface.

The figure is trailing a rope or line of some kind. At the end I see a human-sized bundle being dragged through the opening before he releases the line again. He hovers here for a moment, motionless, and then grabs the flashlight and shines the beam straight up at me.

'Oh my God,' I groan, still floating here on the surface, waiting for the figure as it swims up through the tank towards me. 'What an idiot I've been ...'

CHAPTER 64

'Now this is what we call a real clusterfuck,' Harvey gasps once he has moved up beside me inside the tank and removed his mouthpiece. He aims a harpoon and flashlight at me. 'Of epic proportions, am I right?'

I yank out my own mouthpiece and cautiously inhale some air. Oppressive, almost suffocating, it tastes bitter on my tongue. 'My friend,' I say, spitting in an attempt to get rid of the acrid flavour.

Our voices sound hollow when we speak. Behind Harvey, air bubbles rise from the bottom, bursting in the confined air pocket in which we are located. Immediately afterwards, the dead body of the woman without a face appears on the surface at his back. He has tied her to a line knotted around her waist. Her torso and the hair on the back of her head are visible; the rest hangs underneath, dangling in the murky water.

'So it was you who came to take her from the lighthouse that first night I was there.'

Harvey turns round and shines the flashlight on the corpse floating on the surface behind him. I can just make out a hood of short black hair and shreds of skin swirling around her skull. 'Yes,' he says, turning back again. 'For a minute or two I was actually afraid you were going to jump in after us.'

'Who is she?'

'I think she's called Elena. From Archangel, Murmansk, or one of those cities there.'

'A sex worker?'

'Yes.'

'What was she doing on board the trawler?'

'Money, money, money.' Harvey tries out one of his self-confident grins. This time he fails. The anxiety and stress are too close to the surface, and the corners of his mouth refuse to obey, instead simply quivering faintly without budging an inch. In the end he says: 'I have some bedsits in the city, six bedsits that I rent out to women from Russia who come here to work for a few months in the year.'

'Why could no one find her?'

'Well, here's where it gets a bit tricky.' I am aware of my fingertips and toes growing numb with cold. Harvey too is white around the lips and eyes. 'Arkady, the trawler captain, and I have this arrangement. He transports a few things for me from Russia, plus he brings the occasional girl to do some work in the city.'

'What things?'

'Sex, drugs and rock 'n' roll,' Harvey answers. 'Booze, amphetamines, steroids – you know, things that help to pay the bills. The trawler went down so fast, the crew just had to jump on board the lifeboats. It wasn't until they'd left the trawler that they realised they'd forgotten the girl.'

'How can you forget something like that?'

Harvey shrugs. 'Those Russians. They drink, take dope and keep going, no matter whether there's a fucking storm or not. Maybe she too was high as a kite. Who knows?'

'Then Rasmus found the woman.'

Harvey nods. 'Wrong place at the wrong time, man. That's all. Exactly like you.'

I remain mute and he goes on: 'The storm that took the trawler seemed to go on for ever. A nor'wester – they blow directly through here, and make it almost impossible to dive or manoeuvre the crane on your own in this particular spot, just beyond the tip of the island. So I had to wait. And when the weather finally relented and I headed out here with the catamaran to haul the stuff out, who do I find anchored right above the wreck in his RIB? Yes, the Danish guy. He'd already been down twice and dragged some of my packages as well as the woman up from the depths. I didn't know what to do, and then all of a sudden he was standing there with his back to me, and I hit out. He just toppled over, plunged overboard and lay there without moving a muscle.'

'He didn't die from the blow to his head,' I tell him, struggling to stay afloat. 'Rasmus drowned.'

Harvey nods silently.

'You shouldn't have tied them together with the cable ties.' We can hear water sloshing against the walls every time the currents outside rock the boat.

'My mistake,' he says, with a sigh, and shakes off a cold shudder as another noise sounds from somewhere outside the tank. 'I brought them both with me and sank them under the water at one of the mussel poles out at my farm. They must have become separated when the poles came adrift during the storm and were carried off by the tide. My God, I searched high and low for them.'

'The local police chief and his sergeant never reached the lighthouse that evening, did they?'

Harvey shakes his head, as if trying to peel off images from his subconscious. 'Bjørkang phoned and told me you had found her out by the lighthouse. I realised as soon as anyone found the Danish guy as well, and saw that they'd been tied together—'

'So you killed them?'

'I asked them to pick me up before they headed out to the lighthouse, and hit Bjørkang on the back of the head with a lifting block as soon as their boat came alongside at the farm quay.'

'And Arnt?'

'Arnt and I had a semi-professional relationship as a result of our shared interest in hunting and fishing. He knew that I sometimes carried Russian liquor in the containers I brought back to the city. Arnt realised what we had to do, as soon as I'd managed to calm him down. I told him about Rasmus and about Elena and convinced him there was no way out, that we had to hide Bjørkang's body in one of the tanks. We could say that their boat had foundered in the storm, and that Arnt had made it to dry land, but that Bjørkang had vanished into the waves.'

'Arnt would never have been able to live with what you'd done.' My thighs are smarting with pain after all the exertion involved in staying afloat, and I carefully jiggle loose a couple of the lead sinkers under the water and let them go, without taking my eyes off Harvey and his harpoon.

At first Harvey says nothing, and just nods faintly to himself before starting to talk again – a bit more quietly now. 'I saw him in front of me in the tank opening after we'd dumped Bjørkang's body. Before I had the chance to reconsider, I had pushed him all the way inside and closed the hatch.'

'You let him freeze to death down here in the dark.'

'There was no other way. No other way!' Harvey raises his voice. The reverberation causes more rust to loosen and rain down over us.

'Harvey the Merciful,' I spit out. 'When did you decide to get me mixed up in all this?'

'The perfect scapegoat,' he hisses. Harvey's face is suddenly greyer, either because of the cold and the oppressive, foul air, or perhaps there is something else: recognition of what circumstances are transforming him into. 'I brought Bjørkang's cap out to the lighthouse the next day and made sure the police found it and the bloodstains in the bar. Mostly just to see what would happen.'

'Oh, a police cap,' I say, forcing out a burst of unconvincing laughter. 'You knew that would get the bloodhounds moving, I guess.'

Harvey gives me a probing look as I laugh again. 'Murderers and the stories they tell, eh?' I shake my head, and continue when Harvey makes no comment. 'Do you see how difficult it is to admit the simplest, most insignificant things, even to yourself? You didn't bring a blood-soaked police cap at random, or just to see what would happen, as you say. That's a lie. One you are telling to protect yourself from ... yourself.'

'Really?' Harvey sniffs dismissively. 'How riveting.'

'The human brain is programmed to alter or adjust reality sometimes – an ancient defence mechanism, a ghost in the machine, intended to help us handle and work through trauma and agonising sense impressions. You tell yourself an alternative story, in which you distort details to make them fit your own self-image.' I feel the urge to laugh, even though my teeth are chattering. 'That's why you use language like "suddenly, before I knew it, he just fell," and so on. That's pure bullshit, Harvey. The truth is that you staged a game with me, a scenario you were working on as early as that morning when you took me out to the lighthouse, when you told me that story from your childhood about the child crying out in the swamplands where your family had a cabin. I'm not saying that you'd planned in detail what the outcome of this scenario would be, but you knew how far you were willing to go, as soon as you worked out what kind of person I was—'

'What? Substance-dependent?' He flings out his arms – the flashlight is pointing one way and the harpoon another. 'A drug addict?' For a split second I consider lunging at him in an effort to grab the harpoon, but the distance is too great, and in a flash they're back again, the torch and the harpoon, aimed straight at my chest.

'Among other things,' I continue calmly. 'Nor do I believe that Bjørkang phoned you that night after I'd found the woman. That's another one of these falsifications that your brain has come up with. Maybe Arnt, but not Bjørkang.'

'Does it make any difference?'

'Of course. Not for me, but for you, Harvey. Arnt was jittery, maybe even scared, when he called me. Probably he realised that the woman's body came from that trawler. Maybe he even suspected you of having something to do with Rasmus's disappearance. I think he wanted to tell me about it, but then he chose instead to rely on you and hung up. After that he phoned you and you persuaded him to come out to the farm where you were waiting, ready to kill. No chance event, just two more cold-blooded, premeditated murders.'

'Lies,' Harvey snarls.

'No one believes this *wrong place at the wrong time* crap of yours,' I add. 'Just as we're both absolutely aware that you didn't turn up here with a harpoon because you were in the area totally by chance and wondered whether we might dive for catfish together. You're a calculating killer who's ready to destroy human lives ... and you're here to commit another murder.'

A loud crack sounds through the hull, followed by a shower of rust flakes drizzling down on us. The water takes on a reddish sheen where our flashlight beams meet the oily mixture. Harvey looks at me. His eyes are dark and hollow and his mouth half open. It is as if we are both dreading the next scene in this tableau at the bottom of the ocean.

'It couldn't end any other way, Thorkild,' he finally whispers.

'And that elaborate spectacle in the disco? Why couldn't you just kill me the way you had done with the others?'

The water is sloshing against the metal sides of the tank. The ripples cause more rust to flake off from the roof and hail down over us, followed by a series of hollow echoes that force their way through the sunken bones of the boat.

'I had to go out to the farm with Elena again, and had thought to hide her in the old sheep shelter until things had blown over. After all, I couldn't have her lying in the elk freezer in the garage all through the winter either,' Harvey says when the hammering and rusty rain have ceased. 'I heard you in the bar: you were talking on the phone as you struggled to down enough pills and alcohol to do the job properly. I decided to help you on your way. Hold one last party for you and Elena. God knows, you both needed one.'

At last his laughter is authentic. The noise rebounds from the walls, sending more flurries of rust down on us while the cold courses through my body. 'I stood watching you both from beside the billiard tables. My God, you were totally gone, man. Wasted. Worst of all,' Harvey continues, shaking his head in mock despair, 'was how you managed to jump into the sea and land on the only piece of flotsam for miles around. It's quite a feat to fail so fundamentally.'

With a shrug, I try to force a facial expression to convey contempt: 'What can I say?' The gesture brings on a spasm of pain and cold in my cheek. 'Some people are just born lucky.'

Playing it safe, Harvey puts his flashlight aside, leaving it drifting in the murk between us. He lifts his harpoon and aims it for my chest. 'Thorkild,' he whispers hoarsely. 'Your luck is about to run out …'

CHAPTER 65

Elena's corpse is floating on the surface two or three metres behind Harvey. The water ripples around her body and hair as the link between her and Harvey tightens.

'So you're thinking of shutting me in here too, together with these other bodies of yours? Your own private graveyard? Some killers keep charnel houses for their victims, so that they own them eternally. I spoke to several of that sort while I was in the States.'

'I landed in a fucked-up situation, that's all. When all you can do is either accept it or else throw in your hand. A man will fight back when he's pushed into a corner.' He flashes me an oblique look before rounding off: 'Most men, at least.'

The way our voices echo around the hull, the sharp reverberation and creaking from the vessel, the cold, the rusty rain and the oppressive air invading the walls of my lungs, these all make me nauseous and dizzy.

'I've heard many confidences from murderers and criminals, Harvey,' I say in an effort to regain some sort of control over the conversation. 'Too many. I know all the words, the mechanisms they use to justify their actions, both to themselves and the rest of the world. I also know it doesn't matter what you say. The ghosts you're trying to conceal down here on the seabed, they won't go away. They're going to haunt you for the rest of your life.'

Harvey looks at me. This time he is not even trying to smile.

'There are two kinds of murderer, did you know that? No, probably not. It doesn't matter how many people they kill – that's not what defines which of the two categories they fall into. Do you know what it is? The key factor that distinguishes the two types from each other?'

'No,' Harvey answers, still with the harpoon aimed at my chest while I speak. I'm so cold now that it is a trial to talk, but nevertheless I press on, forcing it out between chattering teeth and ferocious muscle convulsions. Simply because I know that when this conversation is over and everything has been said, what awaits is the cold steel of Harvey's harpoon.

'What distinguishes the two,' I press on, 'is that most murderers are weighed down by guilt over what they have done. They will spend time after each episode trying to forget, suppress and hide. But there are some, a special race of killers, who are different. They are reflected in their crimes. People who collect dead bodies – in fact, they are their trophies. The yardstick for all their achievements. But you, Harvey. You're no trophy hunter, are you?'

Harvey shakes his head gently without speaking. 'It harms you, what you have done. You've already embarked on the painful journey – I can see that in your face.'

Harvey is totally still now: his shoulders have dropped and the hands holding the harpoon are swaying on the surface of the water.

'You just don't know it yourself,' I labour the point. 'Because you are still in the midst of the whirlwind – adrenalin and shock, after all you've been through recently, are all boiling in there, and you're acting in panic. Hitting out at everything and everyone that reminds you of what you've done, because you still believe that this is something you can wash off, something you can escape from.'

I inch a cautious fraction closer to where Harvey is floating, talking all the while and stealing forward on this razor-thin knife-edge. 'You fucked up,' I whisper. 'We're on the ocean floor on board a shipwreck that you've filled with dead bodies. You could have filled every single tank in this boat with your skeletons, and that would still not be enough. You can never escape from this. Kripos is in town, and they're going to check mobile-phone traffic, see who has talked to whom, and when. Soon their mini-submarine will come to trawl the seabed and find this wreck.'

'It can still work out,' he answers weakly.

'Stop all this bullshit and look around!' I splutter, and then am overtaken by a violent coughing fit. Splashing around, I thresh my hands as my body is racked with the strain of breathing. 'You're floating,' I gasp when I catch sight of the tip of the harpoon as it breaks through the water between us. 'You're floating around here in this tank, inside a wreck on the bottom of the sea, with a corpse trailing after you on a fishing line, and then you tell me that you still believe everything can be OK again? That you, Merethe and your little boy can continue just like before? You've trapped yourself in a delusion, Harvey – I know that. I've been there myself.'

'You're the one who's deluded,' Harvey replies coldly, tightening his grip on the harpoon. 'Who believes that this scenario can have more than one ending.'

I know Harvey is right. But all I can do is plough on, moving in ever-decreasing circles as I search for a way in.

'They will turn up again, those ghosts of yours.' I stop advancing. 'When you're alone, when you're eating dinner with your wife, or when you're putting your son to bed at night. Then they'll be there. You will have to share them with the people you love. With Merethe and the boy.'

'They're never going to know about it.'

'Your wife sees dead people. It's her job. And you believe in all that stuff yourself, you already told me that.'

Harvey blinks hard several times over. 'They're never going to know about it,' he repeats mechanically.

'What about your son? What if he possesses that ability too? What if he's the one the ghosts come to when they don't get what they want from you?'

'Wh-what?' Harvey reels back through the water, forced to spread his hands to keep his balance.

'I saw her, Harvey,' I say, pointing at the greyish-black lump of flesh hanging behind him, suspended from a fishing line. 'Elena. In your wife's eyes. I heard her screams.' I'm no more than a metre distant now, and drop my voice: 'Don't you get it?' I say, in a milder tone. 'That I know. I know what is waiting, that's why I'm telling you this. You can't escape this. It has already locked itself deep inside you and will never leave you. All you can do is to come forward, admit what you've done, and take responsibility for it. For your sake, and for Merethe. And for your boy.'

Harvey is at a complete standstill again. Behind him the water spills over the dead body as it bumps against one of the tank's inside walls. It looks as if he is thinking, mulling things over and sorting out the chaos inside him. In an interview situation, we try to find our way towards a crossroads such as this, at which point the subject is forced to choose

whether he can continue with the lie or change his statement and admit what he has done, let go of the falsehoods
and take personal responsibility. All the same, I am pain-
fully aware that the consequences will not be the same here
on the ocean bed, with a man who has already committed
several murders, and who holds a harpoon in his hands, if
the outcome doesn't turn out as I wish.

Finally Harvey directs his gaze up at me again. I can
already see in his eyes that he has made up his mind. Harvey
has not decided to admit anything, or to take anything seriously at all. He is not the type. 'Bloody hell,' he whispers
huskily, raising the harpoon once more, 'all you had to do
was jump in the sea.'

I'm all set to wheel round and dive down into the water
in an attempt to escape, but Harvey has already shot the
bolt. I can feel it enter straight through my hand and carry
further, passing my breastbone at the moment the force
propels me back.

'I'm just helping you do what you couldn't manage for
yourself,' I hear Harvey shout from somewhere in the
darkness. The pains in my chest are so acute that I want
to scream, but each time I open my mouth it fills with
seawater. The harpoon bolt has pinned my hand to my
chest, and I can feel a stinging, piercing wound in my
armpit.

'Was that a hit or a miss?'

I glimpse Harvey on the other side of the tank, holding
a red cord in his hand and dragging it behind him. 'Found
you,' he triumphs, tensing the cord so that it stretches out
between us. 'And that's a hit!' Harvey exclaims before pulling the cord with all his might.

I scream as my body judders forward and is yanked towards Harvey. A burning pain spreads from my armpit all the way through my chest, as if his heaving is tearing a chunk of flesh away from my torso. I twist my head as Harvey drags me towards him, and I get a brief glimpse of the arrowhead sticking out behind my armpit.

I try to check my forward progress with my free hand but it's no use. It comes to me that I have a diver's knife fastened to my hip, and I use my free hand to search for it, all the time struggling to keep my face above water.

The dead body behind Harvey gives another lurch forward as Harvey pulls himself even closer to me. He is reeling me in with the harpoon cord, as if I were a fish.

Harvey is only a few metres away when I finally feel the shaft of the diver's knife. I start to pull and tug in an attempt to free it, while Harvey hauls in the line so that I'm dragged head-first and now face him.

The searing pain in my chest stops all of a sudden when I finally release the knife and cut the harpoon cord. Floundering, I pull with my free hand and my feet until I have shifted my body into an upright position again.

'And what are you going do with that little thing?' Harvey asks when he catches sight of the knife I am holding in front of me. I know there is no chance I can do anything at all as long as I can use only one hand, even with a knife in that hand. Harvey releases the severed cord and pulls up the harpoon again, pointing the weapon straight at me. I can see through the light shed by his torch that he is taking out another bolt from a holster on his thigh and inserting it in the nozzle. Then he starts to pump the weapon.

I get ready for take-off, launching myself backwards and kicking out to flee. In front of me I see Harvey following through the water.

'What's the matter, why are you trying to get away? I thought you wanted to die?' Behind him the line reels out again so that the corpse is dragged through the water in his wake.

I swim a few more strokes before giving up. My legs are so cold and the pains in my chest so severe that I'm gasping for air. I struggle to keep myself afloat as I cough and spit out bile.

Harvey stops a few metres away from me to take aim. He shines the flashlight tucked in under one elbow, as he uses the other arm to guide the harpoon.

'Go to hell!' I bellow, feeling my gullet fill with blood. I lift the hand with the knife out of the water and hold it in front of me in a pathetic attempt to shield myself from what is coming.

'I'm sorry, Thorkild.' Harvey raises the harpoon out of the water and aims for my chest. From behind him comes a loud splash. I can see that Elena's body has drifted all the way across to him and is practically nudging his back. He notices it too, and turns to look.

The dead body seems to be tangled in the line between them, and when Harvey turns round he tugs the corpse even closer, so that she suddenly rises from the water and falls forward on to him.

'Aargh,' he howls, and fires the harpoon in panic. I see the bolt go straight through Elena's stomach and out the other side. Harvey drops the harpoon as he continues to revolve. He tries to shove the body away, sawing his hands and pulling back. Instead of wrenching free, the cords in which they are tangled tighten and their bodies spin towards each other, closer and closer, while Harvey's screams echo around the tank.

Slender rays of light shoot out from Harvey's flashlight, caught in the shapeless, shadowy mass of arms and legs in front of me. Harvey thrashes his arms to strike out and lift the dead body off him. From time to time I hear sporadic gasps and gurgling, guttural sounds, before a rasping roar echoes through the steel tank. Immediately afterwards, the two bodies disappear underwater and are gone.

Chapter 66

The occasional air bubble breaks the surface where Harvey and the dead woman had recently been. Somewhere below me I can make out a couple of pale rays of light, like rabbit holes in the darkness, though I can see no trace of either Harvey or Elena.

My body is heavy, and a numb sensation is spreading from my chest where the arrowhead is lodged. My legs push against the surface of the water, while my stomach and thighs, where the rest of the lead sinkers are located, drag me down. Harsh air sears my lungs, and I have a violent feeling of choking with every intake of breath.

I use my free hand to search for my mouthpiece to connect to the air tanks, as centrifugal force spins my body to one side. Floundering, I pull with my free hand and kick out with my legs in an attempt to right my body. The exercise sends shooting pains through my chest that radiate out into my armpit.

After the third round trip, I suddenly feel something soft on my fingertips. This time I remain prostrate in the water, upside down, as I try to hold my breath and refrain from moving, at the same time fumbling for the mouthpiece tube. My lungs are about to explode when at last I get hold of the rubber mouthpiece and can insert it into my mouth.

These exertions have drained me of strength. I am about to lose all feeling in my legs and am aware of my body being sucked down. It strikes me that I must now shed the remaining sinkers around my thighs, or I'll plummet to the bottom and stay there.

It has grown harder already to breathe through the mouthpiece, either because it is on the point of running out of air from the tank, or because the arrow through my chest has pierced a lung. The instrument panel showing how much air I have left is attached to the hand pinned to my chest by the harpoon. I therefore have no chance of checking how much air remains. I cannot connect the spare tank with only one hand.

Below me I can still see light from Harvey's torch. The rabbit holes are gone and there is only a single star-shaped point of light far down there in the darkness. During my exertions, I have lost my diver's mask, and the LED light on it must have been damaged. The yellow glow down there is the only source of light I can discern. I try to focus my eyes, aware of the urge to sleep suffusing my body – stronger, more urgent and insistent with each movement.

The light below me changes colour from deep yellow to a whiter, clearer light each time I look at it. After a while, I don't even manage to remain in an upright position. My body keels over, and my legs float up so that I'm lying on my side in the water. This time I'm unable to turn back, and instead just hang there, switching in and out of consciousness, as my body drifts on the surface of the water.

I am surrounded by silence, with no idea whether I am above or below the water. I have lost all sensation in my fingertips and the only thing I do feel is the pain in my chest and armpit. I try to move my free hand and haul myself round again. My lips are numb and I no longer feel my mouthpiece even when I bite into the rubber.

My eyes snap open when I hear a loud crack from the boat, and a pounding rumble rises through the metal construction, followed by a shower of fine rusty rain. All at once my face is above water again, and I can see a shadow, more compact and substantial than the rest of the darkness, almost as if someone is hanging there on the roof above me, and at the same instant I am aware of the taste of perfume on my tongue.

I blink repeatedly, as if to wash away the deadly fatigue and the dirty, rust-infected salt water, as the shadow continues to sink towards me.

'Who are you?' I ask, though my lips do not move, and I simply go on drifting, rolling round and in and out of water and consciousness.

No one answers. An intense heat spreads through my body as the shadow drops down from the roof and passes through me. I struggle and strike out with my free hand until I have turned myself over, towards the shadow that is now below me. I try to grab hold of it, pulling it and the heat back. Instead I start to sink.

I am unsure whether I actually move or whether what I see is merely a fag-end of myself, a fine layer of transparent ectoplasm flowing out of the dying body to combine with the seawater. Nevertheless I continue, kicking, scratching and dragging my body through the rusty rain as I bite into my mouthpiece as hard as I possibly can.

Somewhere ahead of me, the yellow light appears again. In a brief glimpse I catch sight of Harvey and Elena in the glimmer from the weak rays of the flashlight between the two dead bodies, tied together by Harvey's line. His eyes are wide open and staring terror-stricken into the gloom as their bodies dance around each other, following the rhythm of the seawater. Suddenly her head glides up so that they are facing each other, the black hair on the back of her head fanning out behind her, and the bodies continue their downward drift until the light between them is gone.

I see the opening in the tank ahead of me, not far from Arnt's back, now lying on the bottom, and turned away from me. I don't see Bjørkang's corpse, and soon I pass through the hatchway and emerge into the place where the seabed is covered with rotten fish corpses. The shadow moves on, up past the hull of the wreck at a slow pace. The shadow is just beyond my reach the entire time. Sometimes it spins round itself, and other times it seems as if the whole shadow is pulsating, taking on a glowing black hue and pointing the way out of the murky fog around the wreck.

We follow the path of dead fish, past the severed wheelhouse towards a pale blue glimmer above our heads. Eventually, as the colour grows stronger, I also begin to kick harder as I claw with my hand to climb faster. I tear the mouthpiece out by accident, but hardly notice it. Every fibre of my being is focused on the blue colour and the heat it discharges.

I break the surface with a hoarse and gurgling guttural scream directed at the sky above me. Seawater streams from my nose and mouth as I lie on the surface of the sea, shouting for joy. The sky is still bright, even though it is humid outside: this semi-darkness is a thousand times brighter than what I have just swum out of. The shadow is gone, leaving only a vague taste of perfume lingering on the tip of my tongue.

In front of me I see Johannes's boat and Harvey's catamaran, moored together. I see no sign of Johannes as I draw closer to the boats and embark on the strenuous process of clambering up the ladder and on board.

Only when I clear the rail do I see him. Johannes is lying at the bottom of the boat between the steering console and the gunwale, a broad gash visible on his head, just above one ear. Below him, bloodstained seawater splashes.

'Johannes,' I croak, as I haul myself all the way across to the seat installation. I lean over the back of the bench, coughing, belching and gasping for air. 'Johannes!' I spit out more blood, more saltwater, and sink to my knees in front of him.

I place two fingers on his carotid artery and feel a weak pulse. 'Don't die, for Christ's sake,' I beg him, before crawling round on the bottom of the boat searching for my belongings. In the end I find my mobile phone. Edging my way over to Johannes again, I lean towards the gunwale, using the side that was not harpooned, as my trembling fingers key in the police emergency number. As soon as I'm finished, I put the phone on my lap and take hold of Johannes's hand.

After a while, I feel a faint vibration on my thighs. Somewhere above me, a gull is screeching. I open my eyes to see that my mobile is ringing: an unknown Stavanger number. I let go Johannes's hand and take the call.

'Yes, hello?' I groan as I struggle to press the mobile against my ear.

'Thorkild Aske?'

'I think so,' I whisper hoarsely.

'Hi, Thorkild. This is Iljana from the unemployment service in Stavanger. Have you a few minutes to spare just now?'

'I hope so,' I say, coughing into the receiver. 'I sincerely hope so.'

'You paid us a visit last week and signed on as a job applicant, isn't that right? Well, I'm calling you today to tell you that I've already arranged a job interview for you at an employment agency here in the city. It's a large telecoms company that needs more staff in their customer-service department out at Forus. Doesn't that sound exciting?'

Letting the phone slip through my fingers, I lean my head against the gunwale and pull Johannes's hand towards me. 'Bloody hell,' I snuffle, squeezing his hand in mine. 'I'm never going to make it on time for that interview at Police Headquarters ...'

THURSDAY

Chapter 67

I am startled when the sit-up mechanism on the bed is switched on and I feel my upper body being slowly raised. I try to turn over into another position, but am instantly aware of a pain shooting through my chest.

'It was you, wasn't it?' asks a stocky man, towering over me like a bear at the edge of the bed. 'You've turned over in your sleep,' he says accusingly, placing a cautious hand on my shoulder.

'Where am I?' I ask groggily, trying to cough up some memory of the circumstances that brought me here.

'You're in the intensive care unit at Tromsø University Hospital,' the man tells me. 'Don't you recall us taking you off the respirator this morning?'

I shake my head warily.

'Here,' the man continues, easing my shoulder forward as he pushes a pillow behind my back.

'What are you doing?' I resist until a stabbing pain knifes from my chest, sending me into paroxysms of coughing that make the whole bed shake.

'Take it easy,' the man protests. 'I'm just trying to help you. You've rolled over on to your back while you were sleeping,' he adds. 'You have to lie on your side – don't you remember we talked about that earlier today?'

This time I don't try to fight him, and the doctor manages to wedge the pillow so that I'm lying almost entirely on my left side.

'By the way, I should pass on regards from Dr Berg at the trauma reception unit,' the man continues, skirting round to the other side of the bed. 'He said you were more dead than alive when they admitted you yesterday afternoon. Both you and your friend.' He pauses momentarily before ploughing on: 'He also asked me to tell you they don't want to see you down there again, and that you need to have a long talk with someone, preferably a professional, about the direction your life has taken.'

'Yes,' I groan, 'I understand. So ...' I try to moisten my lips with the tip of my tongue before the doctor gives me a glass of water from the bedside table. 'That other guy,' I begin. 'The one I came in with. Where is he?'

'He's here, too, in another room. Head injury, lost a lot of blood, slight hypothermia, but he's stable. These old guys, they can stand a knock.'

I peer down at my ribcage where a big compress nestles among several electrodes connected to a screen that monitors my heartbeat. On my right wrist, where the harpoon entered, I have a plaster cast, and one of my fingers is attached to an oximeter. 'What's the damage? It feels like more than just a few scratches on the varnish this time.'

'Well,' the doctor begins, taking a deep breath. 'The harpoon bolt fractured the third metacarpal bone and damaged the fourth when it entered your hand, hence the plaster cast. It looks fine – we had to realign the bones before we could cast your hand in plaster. There's still some potential for minor nerve or tendon damage, but it looks OK at present. The harpoon penetrated further in between two of your ribs and on through the *lobus medius*, that is the middle lobe of the right lung, and out through your armpit. That's why we've inserted a thoracic catheter to prevent your chest cavity filling with blood.'

'Brilliant,' I comment unconvincingly.

'You've been incredibly fortunate, Thorkild. The fact that the harpoon bolt missed all of your nerve centres and the submandibular ganglion, and only grazed the major blood vessel that delivers blood to your whole arm, is practically a miracle. There's still some risk of after-effects, such as lack of sensation in your right arm, but basically, it looks OK.'

'When can I check out?'

'You will have to stay here with us for a couple of days, just so we can be sure that everything is OK. Apart from that, it's only a matter of enjoying the peace and quiet and trying to relax.' Edging round the bed towards the exit, he stops in the doorway and turns round: 'By the way, I nearly forgot. You have a visitor.'

'Who is it?'

'A policeman. If you like, I can say that—'

'It's all right,' I answer, fumbling for the glass of water again. My mouth feels dry, my tongue swollen.

'Cheers!' Brandishing a paper cup, Gunnar Ore positions himself at the foot of the bed. 'Our very own Hercule fucking Poirot, eh?'

'Piss off,' I riposte, coughing.

'The same to you.' Gunnar Ore drags a chair across to the side of the bed I am facing. 'So, how's the boy doing?' he asks, leaning in towards me.

'As you see.' I pull back the bedclothes and show him the compress and electrodes on my chest. 'Never been better. To cap it all, I've been summoned to a job interview for a thrilling opportunity as a temporary customer-service adviser in the telecoms industry. Rumour has it that they also engage people in permanent posts, if you just roll up your sleeves and work like a slave.'

Gunnar Ore shakes his head. 'You're bloody unbelievable,' he says. 'Who would have known you had it in you, eh? I always thought you were the sort who couldn't stand red tape and liked to talk people to death.' He shakes his head again. 'What a world.' He jumps up again and drags the chair back to where it had been. 'Well, my friend, I just popped in one last time before I head off to the next military camp. You know there are people waiting to talk to you, don't you?'

I nod.

'I just want to say that I've cleared the way a bit for you down there at Police Headquarters. I've spoken to Sverdrup and the boy from Kripos, and told them some things about who you are, or rather, who you once were, so that they're sure to treat you the way you deserve, now that the hullabaloo is on its way.'

'Thanks.'

'No, no, no. It doesn't change anything. But nobody else is going to say that you did any good while you were up here, nobody, and if your name crops up in the newspapers beside mine, well, you know yourself what you can expect.' He puts his coffee cup down on the bedside table, and raps his knuckles on the metal rail that runs along the side of the bed. 'Get well, go away and stay away. OK?'

'We'll talk later, Gunnar.'

'No,' Gunnar Ore says once again. 'No, Thorkild. Let me repeat what I just said for the benefit of the guys in the back row.' He leans all the way into my face and hisses through his teeth: 'Get well, go away and stay away. OK? You're no longer a police officer.' He pulls back again and, standing in front of the bed, spreads out his arms. 'You're a ... a bloody pensioner.'

'Potential customer service adviser with prospects for permanent employment,' I shout after him.

'Oh, go to hell.' Gunnar Ore disappears out through the door and marches along the corridor to the lifts, whistling all the way.

'You too,' I mutter under my breath, grabbing the paper cup of piping hot coffee. 'You too ...'

CHAPTER 68

Anniken Moritzen sounds on edge when she picks up the phone. Glasses and cutlery clink in the background, and I can hear Swedish folk music playing.

'I've been trying to call you all day. I'm leaving soon and—'

'I'm sorry. I've been … busy.'

'Are you still up there?'

'Anniken, I know what happened to Rasmus,' I tell her.

'What?' A door slams at the other end, and the music is turned down. 'What did you say?'

'You were right, of course,' I continue. 'Rasmus was not involved in any mischief. He went out diving, as you said, and found something he wasn't meant to find. It cost him his life.'

'Who did it?'

'That doesn't matter. He's dead.'

Another pause ensues – it is almost as if I can hear the Stavanger rain through the handset. 'Why?' Anniken finally asks. 'Why … did he have to be murdered?'

'We can talk about it more when you come up. I'm afraid I can't meet you at the airport, but tomorrow—'

'I can't go there alone,' she says. 'I have to go out to the lighthouse to collect Rasmus's belongings, and we must—'

'Can't Arne come with you?'

400

'No, he's in Houston and won't be back till late tomorrow evening.'

'Phone Ulf. No matter what, he'll jump on the first plane and come up here the minute he finds out I'm in hospital. Again. I'll meet you both tomorrow, or as soon as I can get out of here.'

I don't want to tell her anything more over the phone. I just want her to know that all the rest of it is over, that I have sorted things out for her so that she can begin to mourn. That she no longer needs to lie awake wishing and hoping for something that is in the past, but she should take time to come to terms with her loss. It is important not to fear that sense of loss, but instead to let it in. Let yourself approach it, rather than run away from it. I could have told her that, said that I know how it feels, but that would not have helped. You have to find it out for yourself. When you're ready.

I am a long way off, inside a dream, in Harvey's company again at the bottom of the sea, when the phone rings.

I rub my eyes in an attempt to erase the fear in Harvey's eyes as he turns towards Elena's corpse and it rises through the water behind him. His desperate screams still echo in my ears when I pick up the phone.

'Hello, Aske here,' I answer.

'You found him,' Arne Villmyr says.

'One more grave.' I sit up in the bed. 'Just as you requested.'

'What I requested,' Arne Villmyr chuckles to himself. 'As if anyone would wish for such a thing?'

'I'm sorry, Arne. I wish—'

'There's nothing to apologise for. You did what I wanted you to do. Don't think I'm reproaching you, not even for what happened to Frei. You were merely the element that came in and changed our lives at that point. The same element that has now given us Rasmus back. The one does not make up for the other, but I'm grateful that he'll soon be coming home again. And that Anniken and I have somewhere to go to. That's all; it's nothing. That's just the way things are now.'

'Do you think she ...' I begin to say, but break off mid-sentence.

'What?' Arne asks. 'Loved you?'

'Yes.'

Arne Villmyr gives a brief burst of joyless laughter. 'Frei found it easy to fall in love. Too easy,' he adds. 'But I think you know the answer to that better than the rest of us, don't you? Only the two of you know what took place on that car journey, and where you were going. I wish I could say something more, but I can't. It's over now. Farewell, Aske.'

CHAPTER 69

The room is in darkness, the only illumination a small light above the sink behind me. I can hear the door of my room creak open before it slides shut again with a quiet click. Immediately afterwards, a stooped figure glides in.

Rolling soundlessly all the way across to the bedside table, she reaches out and takes my hand, drawing it out from beneath the quilt to put it between her two hands and squeeze mine. As soon as I look into her eyes, I relax. I know that gaze all too well: the same expression greets me in the mirror every morning.

Merethe looks at me for a while before picking up her writing pad and pen.

How can anyone live a whole different life parallel to the one shared with someone? is what is written on the pad she holds up to me.

I screw up my eyes, as if to wring out the last image I have of Harvey and Elena, closely intertwined at the bottom of that tank. When I open them again, I see that Merethe is writing something else. 'Harvey thought he had to take that route,' I tell her as she writes. 'For you and your son, but—'

Merethe holds up the notepad again: *I see him in the background whenever I close my eyes. It is as if he doesn't dare to come forward to meet me.*

'And Frei?' I whisper in a rasping voice. 'Do you still see her?'

Merethe glances at me before starting to write once more.

I nod at the word on the pad and my gaze goes upwards, across Merethe's face and on towards the curtains that are almost completely closed. I can just make out the contours of a rounded hilltop beneath rain-filled clouds scudding past in the night sky outside. For a moment I think I perceive something in the midst of those flitting clouds, a faint silvery shimmer hidden behind the chaos of grey and black, followed at once by new clouds rolling past, super-imposed on the old ones.

Soon the entire sky is black.

NOTE ON THE TYPE

The text of this book is set in Linotype Sabon, a typeface named after the type founder, Jacques Sabon. It was designed by Jan Tschichold and jointly developed by Linotype, Monotype and Stempel in response to a need for a typeface to be available in identical form for mechanical hot metal composition and hand composition using foundry type.

Tschichold based his design for Sabon roman on a font engraved by Garamond, and Sabon italic on a font by Granjon. It was first used in 1966 and has proved an enduring modern classic.